Lecture Notes in Computer Science 5962

Commenced Publication in 1973
Founding and Former Series Editors:
Gerhard Goos, Juris Hartmanis, and Jan van Leeuwen

Philippe Palanque Jean Vanderdonckt
Marco Winckler (Eds.)

Human Error, Safety and Systems Development

7th IFIP WG 13.5 Working Conference, HESSD 2009
Brussels, Belgium, September 23-25, 2009
Revised Selected Papers

Springer

Volume Editors

Philippe Palanque
University Paul Sabatier, Institute of Research in Informatics of Toulouse (IRIT)
118, Route de Narbonne, 31062 Toulouse Cedex 9, France
E-mail: palanque@irit.fr

Jean Vanderdonckt
Université catholique de Louvain
Place des Doyens 1, 1348, Louvain-La-Neuve, Belgium
E-mail: jean.vanderdonckt@uclouvain.be

Marco Winckler
University Paul Sabatier, Institute of Research in Informatics of Toulouse (IRIT)
118 Route de Narbonne, 31062 Toulouse Cedex 9, France
E-mail: winckler@irit.fr

Library of Congress Control Number: 2009943657

CR Subject Classification (1998): H.5, J.7, J.2, D.2.2, H.5.2

LNCS Sublibrary: SL 3 – Information Systems and Application, incl. Internet/Web and HCI

ISSN 0302-9743
ISBN-10 3-642-11749-X Springer Berlin Heidelberg New York
ISBN-13 978-3-642-11749-7 Springer Berlin Heidelberg New York

springer.com

Printed in Germany

Typesetting: Camera-ready by author, data conversion by Scientific Publishing Services, Chennai, India
Printed on acid-free paper SPIN: 12987610 06/3180 5 4 3 2 1 0

The original version of the book was revised: The copyright line was incorrect. The Erratum to the book is available at DOI: 10.1007/978-3-642-11750-3_10

Foreword

HESSD 2009 was the 7th IFIP WG 13.5 Working Conference in the series on Human Error, Safety and Systems Development which looks at integration of usability, human factors and human–computer interaction within system development. This edition was jointly organized with the 8th TAMODIA event on Tasks, Models and Diagrams for User Interface Development. There is an obvious synergy between the two previously separated events, as a rigorous, engineering approach to user interface development can help in the prevention of human error and the maintenance of safety in critical interactive systems.

Following the tradition of HESSD events, the papers in these proceedings address the problem of developing systems that support human interaction with complex, safety-critical applications. The last 30 years have seen a significant reduction in the accident rates across many different industries. Given these achievements, why do we need further research in this area? Recent accidents in a range of industries have increased concern over the design, management and control of safety-critical systems. Therefore, any system that involves human lives in its functioning is subject to safety-critical aspects. Contributions such as the one by Holloway and Johnson (2004) report that over 80% of accidents in aeronautics are attributed to human error.

Much recent attention has focused upon the role of human error both in the development and in the operation of complex processes. Since its inception, the IFIP 13.5 Working Group in Human Error, Safety, and System Development has organized a regular workshop that is aimed at providing a forum for practitioners and researchers to discuss leading-edge techniques that can be used to mitigate the impact of human error on safety-critical systems. The intention is to focus the workshop upon techniques that can be easily integrated into existing system engineering practices. With this in mind, we hope to address a number of different themes: techniques for incident and accident analysis; empirical studies of operator behavior in safety-critical systems; observational studies of safety-critical systems; risk assessment techniques for interactive systems; safety-related interface design, development and testing. The WG also encourages papers that cross these boundaries and come from many diverse sectors or domains of human activity.

These include but are not limited to aviation, maritime and the other transportation industries, the healthcare industries, process and power generation, and military application.

This book contains eight revised papers selected from the papers presented during the Working Conference that was held in Brussels, Belgium, September 23–25, 2009. The papers presented there resulted from a peer-review process and each paper received at least four reviews from the Program Committee members.

The keynote speaker, Dr. Andreas Lüdtke, Head of the Human-Centred Design Group at OFFIS Institute for Information Technology, R&D Division Transportation, presented an invited paper entitled: "New Requirements for Modelling how Humans Succeed and Fail in Complex Traffic Scenarios."

We gratefully acknowledge the support of the FP7 HUMAN project that supported the organization of this workshop (http://www.human.aero).

November 2009

Philippe Palanque
Jean Vanderdonckt

Holloway and Johnson (2004) Distribution of Causes in Selected US Aviation Accident Reports Between 1996 and 2003, 22nd International Systems Safety Conference, International Systems Safety Society, Unionville, VA, USA, 2004.

Organization

General Chair

Jean Vanderdonckt	Université catholique de Louvain, Belgium

Program Chair

Philippe Palanque	University Paul Sabatier, France

Program Committee

H.B. Andersen	Risoe, Denmark
R. Bastide	University Toulouse 1, France
R.L. Boring	IRisk & Reliability Analysis & Sandia National Laboratories
G. Boy	EURISCO, France
P. Curzon	Queen Mary & Westfield College, UK
M. Harrison	University of Newcastle, UK
C.M. Holloway	NASA Langley, USA
C. Johnson	University of Glasgow, UK
C. Kolski	Université de Valenciennes, France
F. Koornneef	TU Delft, The Netherlands
P. Landkin	University of Bielefeld, Germany
K. Luyten	University of Hasselt, Belgium
J. Melchior	Université catholique de Louvain, Belgium
D. Navarre	University Toulouse 1 Capitole, France
A.-S. Nyssen	University of Liege, Belgium
P. Palanque	Paul Sabatier University, France
A. Parush	Carleton University, Canada
F. Paternò	ISTI-CNR, Italy
C. Santoro	ISTI-CNR, Italy
S. Steere	Centre National d'Etude Spaciales (CNES), France
B. Strauch	National Transportation Safety Board, USA
G. Szwillus	University of Paderborn, Germany
T. van der Schaaf	T.U. Eindhoven, The Netherlands (TBC)
J. Vanderdonckt	Université catholique de Louvain, Belgium

Local Organization

Jean Vanderdonckt
Josefina Gerrero Garcia
Juan Manuel Gonzalez Calleros

Proceedings Editor

Marco Winckler Paul Sabatier University, France

Registration and Sponsorship

Kênia Sousa Université catholique de Louvain, Belgium

Website

Francisco Martinez Ruiz Université catholique de Louvain, Belgium

Sponsoring Institutions

Working Group 13.5: Human Error, Safety, and System Development
IHCS: Interacting Humans with Computing Systems, University Paul Sabatier
Université catholique de Louvain Belgian Laboratory of Computer–Human Interaction (BCHI)

Table of Contents

Invited Talk

Human Factors in Healthcare Systems

Pilot's Behaviour

Ergonomics and Safety Critical Systems

New Requirements for Modelling How Humans Succeed and Fail in Complex Traffic Scenarios

Andreas Lüdtke

OFFIS Institute for Information Technology,
Escherweg 2, 26121 Oldenburg, Germany
luedtke@offis.de

Abstract. In this text aspects of human decision making in complex traffic environments are described and requirements for cognitive models that shall be used as virtual test pilots or test drivers for new assistance concepts are derived. Assistance systems are an accepted means to support humans in complex traffic environments. There is a growing consensus that cognitive models can be used to test systems from a human factors perspective. The text describes the current state of cognitive architectures and argues that though very relevant achievements have been realized some important characteristics of human decision making have so far been neglected: humans use environment and time dependent heuristics. An extension of the typical cognitive cycle prevalent in extant models is suggested.

Keywords: Human decision making, complexity, cognitive modelling, cognitive engineering.

1 Introduction

Every day we as humans are faced with complex scenarios in which we have to make decisions under time pressure. Most people know how to drive a car and most often we manage to reach our destination without being involved in an accident. Undeniably, traffic situations can be very complex. But we have learned to cope with critical situations and often we react intuitively without much thought. But, on the other side the high number of accidents that are attributed to human error [44] clearly shows the limitations of human behavior. One way to reduce the number of human errors is the introduction of assistance systems, like Flight Management Systems in aircraft and Adaptive Cruise Control in cars.

Air traffic environments like road traffic environments are inherently complex. Though pilots are highly trained professionals human error is also the main contributor in aircraft accidents [4]. Modern aircraft cockpits are already highly automated and assistance systems have in parts succeeded in reducing errors but new error types have emerged [13, 38, 39, 40, 47, 49]. As a consequence it has widely been accepted that automation systems must be developed from a human centred perspective putting the pilots or drivers in the center of all design decisions. Cognitive engineering [16, 7, 48] is a research field that "draws on the knowledge and techniques of cognitive psychology and related disciplines to provide the foundation for principle-driven

The original version of this chapter was revised: The copyright line was incorrect. This has been corrected. The Erratum to this chapter is available at DOI: 10.1007/978-3-642-11750-3_10

P. Palanque, J. Vanderdonckt, and M. Winckler (Eds.): HESSD 2009, LNCS 5962, pp. 1–17, 2010.

design of person-machine systems" [48]. One line of research in this area deals with developing executable models of human behavior that can be used as virtual system testers in simulated environments to predict errors in early phases of design. But the question arises whether the current human models are capable of simulating crucial aspects of human decision making in complex traffic environments.

This text provides a short introduction in human modeling from the perspective of production system architectures (like ACT-R [3], SOAR [50] and CASCaS [25]) and shows how such models can be used in cognitive engineering approaches. Starting from a definition and two examples of complexity characteristics of human behavior will be elaborated based on results from research on Naturalistic Decision Making [22] and driver perception. The central message is that human decision making is based on heuristics that are chosen and applied based on features of the environment and on available time.

The environment and time dependent application of heuristics has so far been neglected in cognitive architectures. In order to capture these aspects human models should incorporate (1) meta-cognitive capabilities to choose an adequate heuristic for a given decision situation and (2) a decision cycle whose quality of results improves as deliberation time increases.

2 Examples of Complex Traffic Situations

In this section two examples of complex decision situations where humans might make erroneous decisions and where potentially assistance systems might provide support will be introduced. The crucial point is that before assistance systems are to be introduced we have to understand how humans make decisions in such scenarios and we have to be sure that with the new systems errors are really prevented and no new errors are introduced.

The first example describes an air traffic situation where pilots have to decide which airport to use (Fig. 1). An aircraft is flying towards its destination airport Frankfurt Main (EDDF). On their way the pilots receive the message that due to snow on the runway the destination airport is temporally closed. Further information is announced without specifying when. The options now for the pilots are either (1) to go ahead to the original airport (EDDF) and to hope that the runway will be cleared quickly, or (2) to divert to the alternate airport Frankfurt Hahn (EDFH) or (3) to request a holding pattern in order to wait for further information on the situation at EDDF. The goals are to avoid delays for the passengers and to maintain safety.

There are several aspects to be taken into account. If the pilots go ahead there might by the possibility that the runway will not be cleared quickly and that in the end they have to divert anyway. This would cause a delay because the aircraft will have to queue behind other aircraft that decided to divert earlier. If they divert a question to be answered is, if a delivery service will still be available which takes the passengers to the original destination, furthermore, if the duty time of the pilots will expire so that they will not be able to fly the aircraft back to the original airport. If they wait for further information there might be the chance that the pilots receive news that in the end the runway is re-opened. On the other hand, there is the chance that it will not be re-opened and a diversion is the only option left after some time of waiting. Will there still be enough fuel for this case?

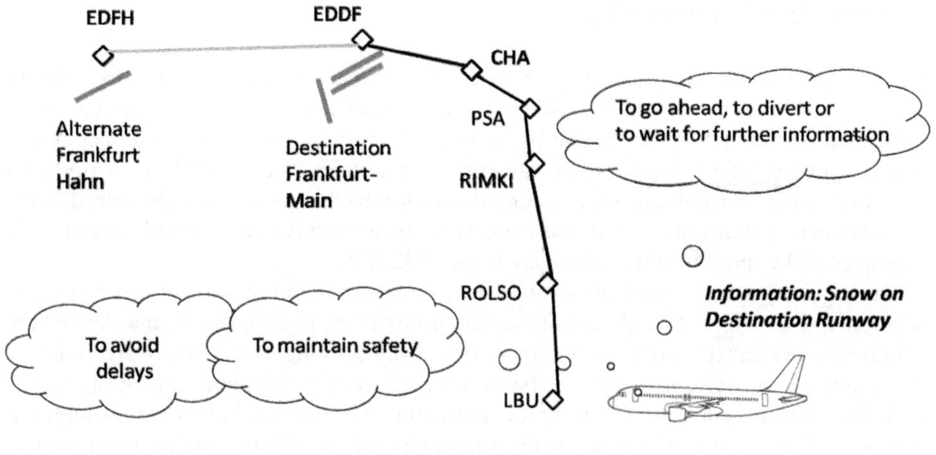

Fig. 1. Complex air traffic scenario

The second example describes a road traffic example in which a car driver has to decide either to stay behind a lead car or to overtake (Fig. 2). If (s)he intends to overtake then (s)he can either let the approaching car pass or not. For these decisions the speed of and distance to the approaching car as well as the lead car have to be assessed. Furthermore, the capabilities of the ego car have to be taken into account. Accident studies have shown that the problem in overtaking scenarios "stems from faulty choices of timing and speed for the overtaking maneuver, not a lack of vehicle control skills as such" [7].

Both examples will be used throughout the text to illustrate characteristics of human decision making in complex traffic scenarios.

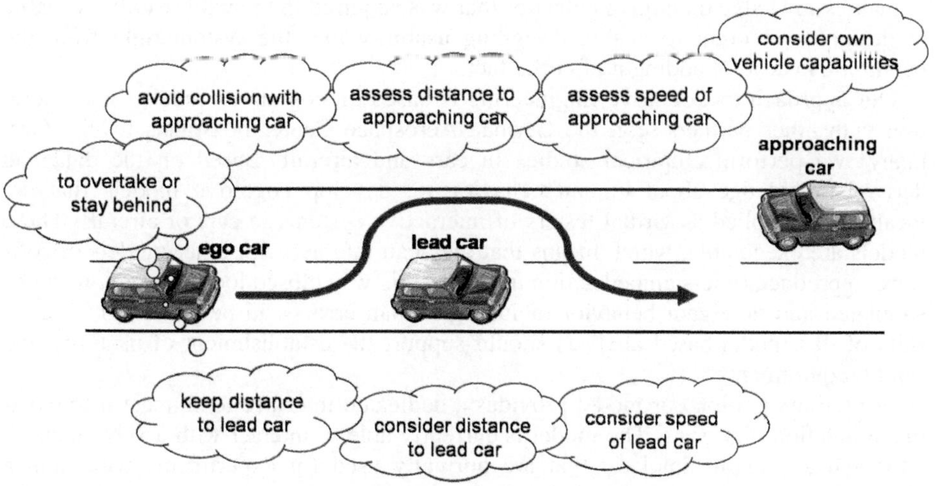

Fig. 2. Complex road traffic scenario

3 Cognitive Engineering

In the design of systems that support humans in complex environments, like the air and road traffic environment described above, characteristics of human behavior have to be understood and should be the basis for all design decisions. Such characteristics include potential human errors. In transportation human error is still the major contributing factor in accidents. One accepted solution to this problem is the introduction of assistance systems in aircraft and cars. Such systems have been introduced but still they need to be more intuitive and easy to use [38, 39].

During design and certification of assistance systems today, human error analysis is perceived as relevant in almost all stages: it has to be proven that human errors are effectively prevented and no new errors or unwanted long-term effects are induced. Nevertheless, the current practice is based on engineering judgment, operational feedback from similar cars or aircraft, and experiments with test users when a prototype is available. Considering the increasing complexity of the traffic environment and of modern assistance systems that are currently researched (e.g. 4D Flight Management Systems in aircraft and Forward Collision Warning in cars) methodological innovations are needed to cope with all possible interactions between human, system and environment. New methods have to be affordable and applicable in early design phases.

Cognitive Engineering is a research field that addresses this issue. Research focuses on methods, techniques and tools to develop intuitive, easy to use, easy to learn, and understandable assistance systems [31]. The field draws on knowledge of cognitive psychology [48] but stresses the point that design and users have to be investigated and understood in "in the wild" [33]. The term "cognition in the wild" has been introduced by Edwin Hutchins [20] and means that natural work environments should be preferred over artificial laboratory settings because human behavior is constrained on the one hand by generic cognitive processes and, equally important, on the other hand by characteristics of the environment. The objective of Cognitive Engineering is to make knowledge on human behavior that was acquired in the wild readily available to designers in order to enable designing usability into the system right from the beginning instead of adding it after the fact.

Our approach to Cognitive Engineering is based on cognitive models. In cooperation with other partners (e.g. the German Aerospace Center in Braunschweig, Germany) we perform empirical studies in cars and aircraft. Based on the data and derived knowledge about human behavior we develop cognitive models that are meant to be applied as virtual testers of interactive systems in cars or aircraft. These models are executable, which means that they can interact with other models or software to produce time-stamped action traces. In this way closed loop interaction can be simulated and emergent behavior including human errors can be predicted. The results of this model-based analysis should support the establishment of usability and safety requirements.

For the integration our model provides a dedicated interface to connect it to existing simulation platforms. The model is currently able to interact with a vehicle simulator and a cockpit simulator that are normally used for experiments with human subjects. The integration with these platforms has got the advantage that the model can interact with the same environment as human subjects. Thus, model data and human data produced in the very same scenarios can be compared for the purpose of model validation. The current status of our aircraft pilot crew model is presented in another article in this book [25].

4 Cognitive Models

The models that are most interesting for Cognitive Engineering are integrated cognitive models. Research on integrated models was proclaimed amongst others by Newell in the early seventies (see e.g. [30]). Newell argued in favor of a unified theory of cognition [29]. At that time and still today (and for a good reason) psychology is divided in several subfields like perception, memory, motivation and decision making in order to focus on clearly defined phenomena that can be investigated in a laboratory setting. The psychology of man is approached in a "divide and conquer" fashion in order to be able to design focused laboratory experiments revealing isolated phenomena of human cognition. Newell [29] suggested to combine the existing knowledge into an integrated model because most tasks, especially real world tasks, involve the interplay of all aspects of human cognition. The interaction with assistance systems involves directing attention to displays and other information sources and perceiving these cues to build up and maintain a mental model of the current situation as a basis for making decisions on how to operate the system in order to achieve current goals.

Integrated cognitive models can be built using cognitive architectures. Cognitive architectures are computational "hypotheses about those aspects of human cognition that are relatively constant over time and relatively independent of task" [36]. They allow to reuse empirically validated cognitive processes and thus they ease the task dependent development of a cognitive model. The architecture integrates mechanisms to explain or predict a set of cognitive phenomena that together contribute to the performance of a task.

A lot of cognitive architectures have been suggested and some have been used to model human behavior in traffic. An overview of cognitive models is provided in [35, 23,18, 14]. The most prominent representatives are ACT-R [3] and SOAR [50]. ACT-R (Atomic Components of Thought Rational) stems from the early HAM (Human Associative Memory) model [2], a model of the human memory. SOAR was motivated by the General Problem Solver [28] a model of human problem solving. These different traditions led to complementary strength and weaknesses. ACT-R has a sophisticated subsymbolic memory mechanism with subsymbolic learning mechanisms enabling simulation of remembering and forgetting. For SOAR, researchers only recently began to incorporate similar mechanisms [6, 32]. One outstanding feature of SOAR is its knowledge processing mechanism allowing to deal with problem solving situations where the model lacks knowledge to derive the next step. In such "impasses" SOAR applies task-independent default heuristics with predefined criteria to evaluate potential solutions. Solutions to impasses are added to the knowledge base by SOAR's universal learning mechanism (chunking).

Both architectures were extended by incorporating perceptual and motor modules of the EPIC architecture (ACT-R/PM [3]), EPIC-SOAR [5]) to be able to interact realistically with simulated environments. EPIC [27] is an architecture that focuses on detailed models of constraints of the human perceptual, and motor activity, knowledge processing is considered with less accuracy. ACT-R and SOAR neglected multi-tasking and thus were criticised for not being capable to model human behaviour in highly dynamic environments like car driving or flying an airplane. Aasman [1] used SOAR to investigate this criticism, by applying SOAR to model approaching and handling of intersections (SOAR-DRIVER). To incorporate multi-tasking, he modelled "highly intersection specific rules" for sequentially switching between tasks like eye-movements, adjust speed, adjust trajectory, attend, and navigate. Contrary to this task-specific approach, Salvucci [37] tried to develop a "general executive" for ACT-R/PM that models task-switching based on dynamic prioritization in a most generic form. His technique is based on timing requirements of goals (start time and delay) and task-independent heuristics for natural pre-emption-points in tasks. He tried to schedule tasks for car control, monitoring, and decision making in lane change manoeuvres.

Further cognitive architectures were motivated by the need to apply human models to the evaluation of human interaction with complex systems (MIDAS (Man-machine Integration Design and Analysis System) [8] and APEX (Architecture for Procedure Execution) [15]. These models focused on multi-tasking capabilities of humans from the very start of their development, but they neglected for example cognitive learning processes. MIDAS and APEX offer several tools for intuitively interpreting and analysing traces of human behaviour.

CASCaS (Cognitive Architecture for Safety Critical Task Simulation) is a cognitive architecture which is developed at the OFFIS Institute for Information Technology [24, 26]. It draws upon similar mechanisms like those in ACT-R and SOAR but extends the state of the art by integrating additional mechanisms to model the cognitive phenomena "learned carelessness", selective attention and attention allocation.

Cognitive architectures provide mechanisms for simulating task independent cognitive processes. In order to simulate performance of a concrete task the architecture has to be complemented with task dependent knowledge. Task knowledge has to be

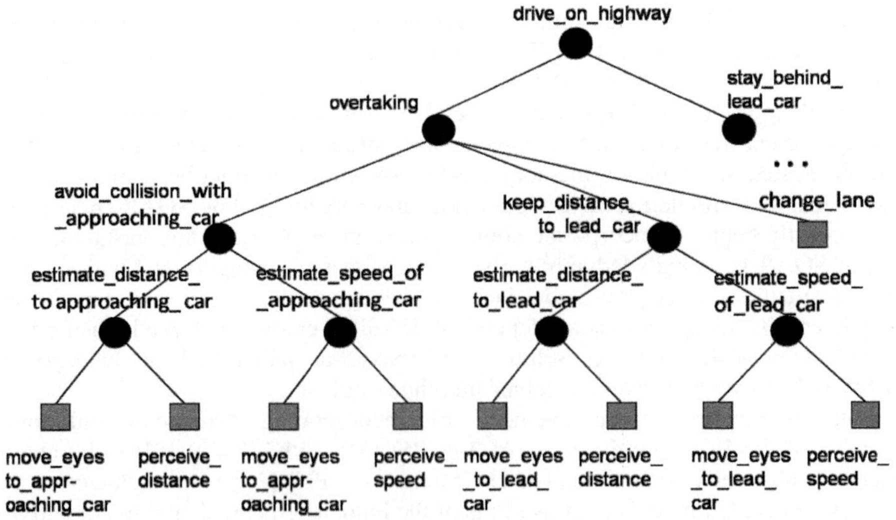

Fig. 3. Task tree for overtaking

modelled in formalisms prescribed by the architecture, e.g. in form of production rules (e.g. ACT-R, SOAR, CASCaS) or scripts (e.g. MIDAS). A common structure behind these formalisms is a hierarchy of goals and subgoals which can be represented as a task tree or task network. In Fig. 3 a task tree for the overtaking manoeuvre in the road traffic example from above is shown. In this tree a top level goal is iteratively decomposed into subgoals until at the bottom concrete driver actions are derived that have to be performed in order to fulfill a goal. The goals as well as actions can be partially ordered.

Every decomposition is either a conjunction or a disjunction. Conjunction means all paths have to be traversed during task performance. Paths may be partially ordered. Within the constraints of this order sequential, concurrent or interleaved traversal is possible. Disjunctions are annotated with conditions (not shown in Fig. 3) that define which paths are possible in a concrete situation. From these possibilities either one or several paths can be traversed. The choices that are not fully constrained by the task tree like sequential/concurrent/ interleaved and exclusive/ inclusive path traversal are defined by the cognitive architecture. In this way the architecture provides an operational semantics for the task tree which is based on a set of psychological phenomena.

Fig. 4 shows a simplified schema of a generic cognitive architecture. It consists of a memory component where the task knowledge is stored, a cognitive processor which retrieves knowledge from memory and derives actions, a percept component which directs attention to objects in the environment and retrieves associated data, and a motor component that manipulates the environment. The interaction of these components during the execution of task knowledge can be described in form of a cognitive cycle as illustrated in Fig. 5 in form of state automata. The cycle starts with the selection of a goal from a goal agenda - the goal agenda holds at any time the set of goals that have to be achieved. Next, new information is received from the percept or the memory components. Based on this data the next branch in the task tree can be chosen which then leads to motor actions (e.g. movements of eyes or hands), memory actions (storing new information) or new goals.

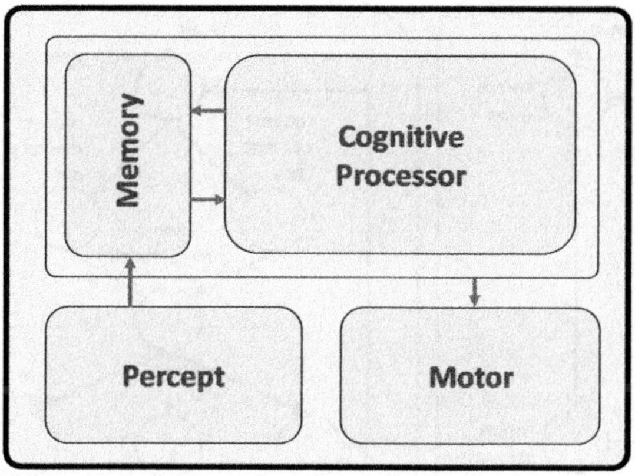

Fig. 4. Generic Cognitive Architecture

In order to illustrate the cognitive cycle the processing of a small (and simplified) part of the decision tree (Fig. 3) shall be explained. For this the task tree first shall be translated into production rules that are the central formalism in production system architectures like ACT-R, SOAR, and CASCaS (see Fig. 6). Let's assume the model's perceptual focus and attention is on the lead car. Fig. 6 illustrates four iterations of the cognitive cycle:

- *Cycle 1*: The currently selected goal is to drive on a highway. The speed of the lead is perceived from the percept component and the ego car speed is retrieved from the memory component. Since the lead car is slower than the ego car it derives a goal to overtake (by selecting rule 1, Fig. 6).
- *Cycle 2*: Overtaking is selected as the next goal and by applying rule 2 the action to move the eyes to the approaching car is derived (in this step no information has to be retrieved or perceived).
- *Cycle 3*: Next the current goal is kept and the action to move the attention to the approaching car is derived (by rule 3) which allows to perceive speed and distance information about the approaching car.
- *Cycle 4*: Again the current goal is kept, information about the approaching car is perceived from the percept component and information about the lead car is retrieved from memory. This information is evaluated and rule is 4 is applied to derive a motor action to change the lane.

This cycle is the basis for cognitive architectures like ACT-R, SOAR and CASCaS. The explicit distinction between moving the eyes and afterwards moving attention separately is a feature that has been introduced by ACT-R and again shows how the cognitive architecture provides a specific operational semantics for task knowledge. The distinction between movements of eye and attention is based on research in visual attention [45, 3] which shows two processes: pre-attentive processes allowing access to features of an object as color, size, motion, etc. and attentive processes allowing access to its identity and more detailed information, e.g. the type of car.

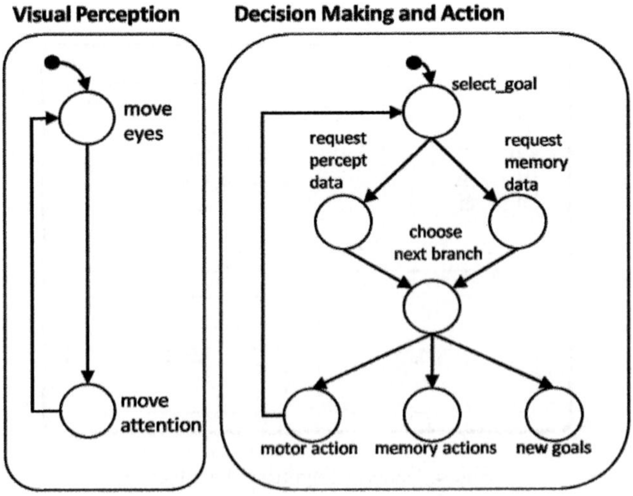

Fig. 5. Typical cognitive cycle

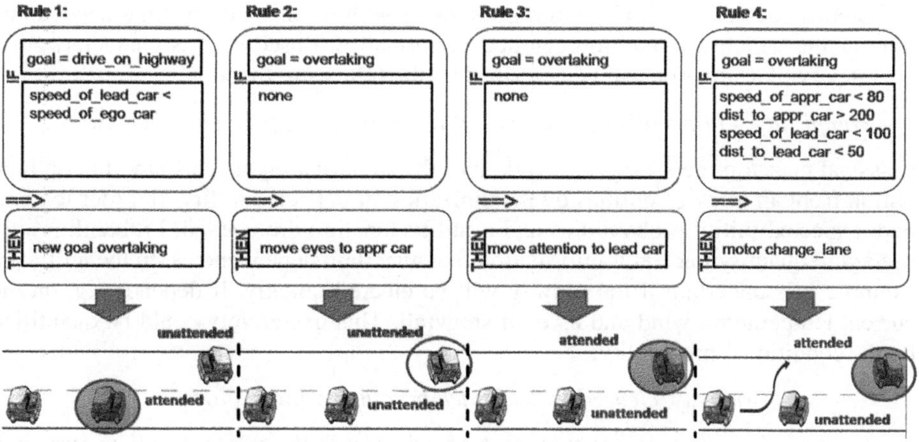

Fig. 6. Examples of rules for the overtaking manoeuvre

In the cognitive cycle described above decision making (if and when to overtake) is modelled as traversing a task tree or network with choice points. The question arises if this concept is adequate to simulate human behaviour in complex dynamic traffic environments. In this paper it is argued that the cognitive cycle has three important shortcomings: (1) processes of visual perception deliver data from the environment independent on the current situation, (2) there is no flexibility with regard to the decision strategy (traversing networks with choice points), and (3) the influence of time pressure is not considered.

These shortcomings simplify some very important characteristics of how humans cope with complexity. One major point is that humans use heuristics for vision and decision making to reduce complexity and to cope with limitations of the human cognitive system. The application of such heuristics is dependent on available time.

5 Decision Making in Complex Air Traffic Scenarios

In this section it will be described how pilots might make decisions in the air traffic scenario introduced above. Before doing so, the concept of complexity shall be further outlined in order to explicate the perspective underlying the decision procedures described below.

The concept of complexity in this text is in line with the definitions given in the field of Naturalistic Decision Making (e.g. [34, 22]). Complexity is viewed as a subjective feature of problem situations. The same situation can be complex for one person but simple for another one. The level of complexity attributed to a situation is highly dependent on the level of experience a person has already acquired with similar situations. Due to experience people are able to apply very efficient decision making heuristics [51]. Nevertheless, it is possible to pinpoint some characteristics of situations that people perceive as complex: conflicting goals, huge number of interdependent variables, continuously evolving situation, time pressure (evolving situations require solution in real-time), criticality (life is at stake), uncertainty (e.g. because of ambiguous cues). Complexity often goes along with a mismatch of time needed and time given, which can lead to degraded performance. Based on this characterization complexity of a situation can be described by the following function:

$$complexity = f(\,problem\_features,\ known\_heuristics,\ applicable\_heuristics\,).$$

Classical decision theory (e.g. [21]) defines decision making as choosing the optimal option from an array of options by maximization of expected utility. In order to compute expected utility probabilities and a utility function are needed. Probabilities are needed to quantify uncertain information like uncertain dependencies. In the air traffic example it is uncertain if the runway will be cleared quickly. It depends e.g. on the current temperature, wind and level of snowfall. This uncertainty could be quantified by the conditional probability:

$$P\,(\,runway\_cleared\_quickly \mid temperature,\ wind,\ snowfall\,).$$

Further probabilistic considerations to be made are: If the pilots decide to wait will their duty time expire in case they have to divert later on? Will there still be enough fuel for a diversion? How long do they have to wait until further information will be available? If they decide to divert will the delivery service for the passengers still be available at time of arrival?

Utilities are needed in order to quantify for all possible situations the level of goal achievement. This has to be done for all goals and for all possible situations. In the air traffic scenario there are mainly two goals: to avoid delays and to maintain safety. The first utility could be defined as hours of delay using the following function:

$$U\!:\ delivery\_service\_still\_available \times expiring\_duty\_time \times diversion \times continue\_to\_original\_airport \times waiting \rightarrow hours\_of\_delay$$

Assuming that each variable is binary (and that the three decision variables are mutually exclusive) the foreseen hours of delay have to be given for 12 situations. Additionally the utility for maintaining safety has to be quantified. In summary, from the perspective of classical decision theory complexity can be defined by the function:

$$complexity = f(\,\#^{1}options,\ \#influence\_factors,\ \#probabilities,\ \#goals,\ \#utilities\,).$$

Classical decision theory was criticized by many researchers as inadequate to describe actual decision making of humans. E.g. Simon [42] stated that the „capacity of the human mind for formulating and solving complex problems is very small compared with the size of the problems whose solution is required for objectively rational behavior in the real world – or even for a reasonable approximation to such objective rationality". He coined the term "Bounded Rationality" [41]. Tversky and Kahnemann [46] described several decision heuristics people use in complex situations to cope with the limits of human decision making. Building on this seminal work the research field Naturalistic Decision Making investigates the way in which people actually make decisions in complex situations [22]. A main point brought up in this field is that proficient decision makers rarely compare among alternatives, instead they assess the nature of the situation and select an action appropriate to it by trading-off accuracy against cost of accuracy based on experience. Experience allows

---

[1] # meaning "number of".

people to exploit the structure of the environment to use “fast and frugal heuristics” [17]. People tend to reduce complexity by adapting behaviour to the environment. Gigerenzer introduced the term “Ecological Rationality”[2] which involves analyzing the structure of environments, tasks, and heuristics, and the match between them. By the use of structured interviews with decision makers several generic decision heuristics have been described [51]. Three of these are Elimination by Aspects, Assumption-Based Reasoning and Recognition-Primed Decision Making. In the sequel, it will be shown how pilots might use these heuristics to make a decision in the air traffic example.

Elimination by Aspects is a procedure that sequentially tests choice options against a number of attributes. The order in which attributes are tested is based on their importance. This heuristic can be applied if one of several options (in our example either to continue to the original airport, to divert to the alternative airport or to wait for further information) must be selected and if an importance ordering of attributes is available. Assuming the following order of attributes *snow_on_runway*, *enough_fuel*, *expiring_duty_time*, *delivery_service* a decision could be done in three steps. (1) There is currently snow on the runway, thus the original airport is ruled out. The remaining options are either to divert or to wait. (2) Because there is enough fuel for both decisions, the second attribute does not reduce the set of options. (3) If a diversion to the alternate is chosen the duty time will expire and there is no chance to fly the passengers to the final destination if the situation has cleared up. Consequently, a diversion is ruled out. Finally, there is only one option left which is to wait. Since a decision has been found the last attribute *delivery_service* is not considered because the strategy is non-compensatory. From the perspective of Elimination by Aspects complexity can be defined as:

$$complexity = f(\#options, \#known\_discriminating\_attributes).$$

The more options the more complex, but complexity is drastically reduced if discriminating attributes are available. The strategy does not necessarily use all attributes but focuses on the more important ones.

In Assumption-based Reasoning assumptions are generated for all unknown variables. For example, the pilots might assume that the runway will not be cleared quickly and that landing on the original airport will thus not be possible during the next hours. This would be a worst case assumption. Consequently, they would decide to divert. Roughly complexity for this heuristic depends on the number of assumptions that have to be made or on the number of unknown variables:

$$complexity = f(\#unknown\_variables).$$

Using Recognition-Primed Decision Making, the third heuristic, people try to recognize the actual situation by comparing it to similar situations experienced in the past. In this way expectations are generated and validated against the current situation. If expectations are met the same decision is taken. For example, the pilots recall a similar situation where there was snow on the original runway and further information were announced. In that situation the temperature was normal, wind was modest. The decision at that time was to wait for further information. Finally, the runway was

[2] Ecological Rationality and Naturalistic Decision Making are very similar but do not follow exactly the same research path. Differences are described in [43].

cleared quickly and the pilots could land. Based on this past situation, the pilots might verify expected attributes like the temperature and wind and if these fit with the past situation they could decide in the same way. The complexity might by defined by:

$$complexity = f(\#known\_similar\_situations, \#expectations).$$

From these three examples of decision procedures, the first conclusion for human decision making in complex scenarios shall be derived:

> *Humans use heuristic decision procedures to reduce the complexity of a situation. The use of heuristics depends on the given information and on the mental organization of knowledge.*

The human cognitive system and the structure of the environment in which that system operates must be considered jointly, not in isolation from one another. The success of heuristics depends on how well they fit with the structure of the environment. In Naturalistic Decision Making cognition is seen as the art of focusing on the relevant and deliberately ignoring the rest.

6 Decision Making in Complex Road Traffic Scenarios

In this section it will be described how car drivers might make decisions in the road traffic scenario introduced above. The description starts, like above, from a normative perspective.

Normatively a car driver has to consider the following information in order to decide if it is safe to overtake or not [19]:

- The Distance Required to Overtake (DRO) as a function of distance to lead car, relative speed and ego vehicle capabilities,
- Time Required to Overtake (TRO) as a function of distance to lead car, relative speed and ego vehicle capabilities,
- Time To Collision with lead car (TTC_{Lead}) as a function of distance to lead car and relative speed,
- Time To Collision of approaching car with DRO (TTC_{DRO}) as a function of speed and distance between DRO and approaching car.

Overtaking is possible if $TRO < TTC_{DRO}$; the safety margin can be computed as $TTC_{DRO} - TRO$.

The problem is that this normative information is not always available. Instead, drivers use visual heuristics [12, 10]. Gray and Regan [19] investigated driver behavior in overtaking scenarios. They identified three strategies for initiating overtaking manoeuvres: (1) Some drivers initiated overtaking when TTC_{DRO} minus TRO exceeded a certain critical temporal margin, (2) others initiated overtaking when the actual distance to the approaching car was greater than a certain critical distance, (3) a thirdgroup of drivers used a dual strategy: they used the distance strategy if the rate of expansion was below recognition threshold and they used the temporal margin strategy if the rate of expansion was above recognition threshold. The rate of expansion is defined based on the angle ϕ which stands for the angular extend of an object measured in radians. The quotient $\delta\phi / \delta t$ is the rate of expansion. It is assumed that peoples' estimation of TTC can be described by the formula $\phi / (\delta\phi / \delta t)$. This formula is an example of an optical invariant which means that it is nearly perfectly correlated with the objective information that shall be measured [9]. Apart from such invariants people also use optical heuristics if invariant information is not available [9]. For example, the rate of expansion (motion information) becomes more impoverished as viewing distance increases. It is assumed that if motion information becomes available drivers use optical invariants like rate of expansion, otherwise drivers use visual heuristics like pictorial depth cues [9]. An example for pictorial depth cues is the size in field or relative size. The use of these cues can sometimes lead to misjudgments. In an experiment DeLucia and Tharanathan [10] found that subjects estimated a large distant object to arrive earlier than a near small object.

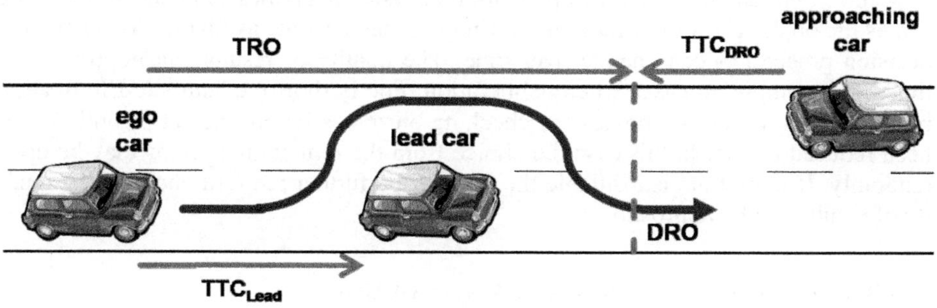

Fig. 7. Normative information for overtaking manoeuvre

Based on this investigation the second conclusion for human decision making in complex scenarios shall be derived:

> *People use visual heuristics to cope with limitations of the human vision system in highly dynamic environments. The use of these heuristics depends on information that is perceivable. If only distance information is available pictorial depth cues are used if motion information becomes available temporal information is used instead.*

Apart from distance to an object further parameters relevant for use of visual heuristics are motion in space, the nature of the current task [11] and visibility.

7 Extending the Typical Cognitive Cycle

Based on the two conclusions derived above two implications for cognitive modeling of human behavior in complex situations will be shown in this section. As a result extensions of the cognitive cycle introduced in Fig. 5 are suggested.

The first implication is that the cognitive cycle needs to be more flexible: The typical cognitive cycle models decision making as traversing a decision tree. In Section 5 it has been shown that people are very flexible in applying decision procedures. In order to model this behavior traversing a task tree should just be one of several other mechanisms for decision making. Additionally meta-cognitive capabilities to choose an adequate heuristic for a given decision situation based on environmental characteristics and available knowledge have to be added to the model. This extension is shown in Fig. 8 as a sub structure for the box decision making and action. There are sub boxes for different decision procedures which all could be further specified by state automata. On top of these a box for meta cognition is added which passes control to the decision procedures.

The second implication is that perception has to be modeled dependent on factors like distance to an object. Visual heuristics are applied in case that optical invariants are not available. This behavior has to be added to the percept component of the model. Based on physical parameter of the current situation it has to be assessed on which cues humans would most likely rely. This is modeled by including different perception mechanisms in the perception box (Fig. 8) that act as filters of incoming information extracting either invariants or different forms of visual heuristics.

A third implication is that the application of visual heuristics can change over time e.g. as the object gets closer motion information may become available. Also heuristic decision procedures can change over time. The quality of results can improve over time. For example, in cases when deliberation time is short the heuristic Elimination by Aspects may stop the process of checking attributes before the set of options has been reduced to one. In this case the choice from the remaining options may be done randomly. If more time is available the set may be further reduced and thus the quality of results can be improved.

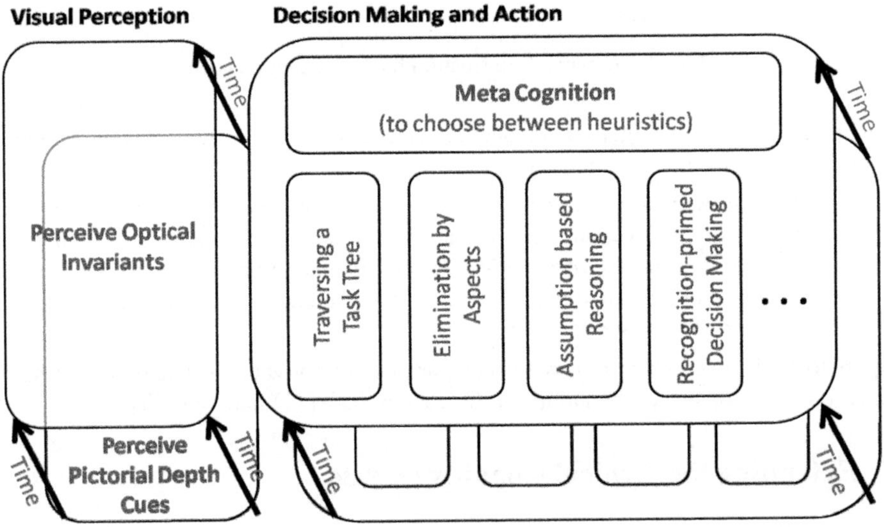

Fig. 8. Extended Cognitive Cycle

As a consequence time should be added as a new dimension to the cognitive cycle (Fig. 8). This new dimension may have two effects:

(1) as time passes the current heuristic could be stopped (e.g. relying on optical depth cues) and another heuristic may be started (e.g. relying on optical invariants),
(2) as time passes the current heuristic may deliver improved results.

8 Summary

In this text the typical cognitive cycle prevalent in cognitive architectures has been illustrated. Human decision making has been described based on two examples from the Aeronautics and Automotive domain. Based on research from Naturalistic Decision Making and visual perception of drivers important characteristics of human behaviour in complex traffic environments have been described. From these characteristics new requirements for cognitive modeling have been derived. The requirements have been introduced in from of extensions of the typical cognitive cycle of cognitive architectures. The text addressed the application of cognitive models as virtual testers of assistance systems in cars and aircraft.

References

1. Aasman, J.: Modelling Driver Behaviour in Soar. KPN Research, Leidschendam (1995)
2. Anderson, J.R., Bower, G.: Human associative memory. Winston & Sons, Washington (1973)
3. Anderson, J.R., Bothell, D., Byrne, M.D., Douglass, S., Lebiere, C., Qin, Y.: An integrated theory of the mind. Psychological Review 111(4), 1036–1060 (2004)
4. Boeing Airplane Safety: Statistical Summary of Commercial Jet Aircraft Accidents: Worldwide Operations, 1959-2005. Boeing Commercial Airplane, Seattle, WA (2006)
5. Chong, R.S., Laird, J.E.: Identifying dual-task executive process knowledge using EPIC-Soar. In: Proceedings of the Nineteenth Annual Conference of the Cognitive Science Society. Lawrence Erlbaum Associates, Hillsdale (1997)
6. Chong, R.: The addition of an activation and decay mechanism to the Soar architecture. In: Proceedings of the 5th International Conference on Cognitive Modeling (2003)
7. Clarke, D.D., Ward, P.J., Jones, J.: Overtaking road accidents: Differences in manoeuvre as a function of driver age. Accident Analysis and Prevention 30, 455–467 (1998)
8. Corker, K.M.: Cognitive models and control: Human and system dynamics in advanced airspace operations. In: Sarter, N., Amalberti, R. (eds.) Cognitive Engineering in the Aviation Domain, pp. 13–42. Lawrence Erlbaum Associates, Mahwah (2000)
9. Cutting, J.E., Wang, R.F., Fluckiger, M., Baumberger, B.: Human heading judgments and object-based motion information. Vision Res. 39, 1079–1105 (1999)
10. DeLucia, P.R., Tharanathan, A.: Effects of optic flow and discrete warnings on deceleration detection during car-following. In: Proceedings of the Human Factors and Ergonomics Society 49th Annual Meeting, pp. 1673–1676. Human Factors and Ergonomics Society, Santa Monica (2005)
11. DeLucia, P.R.: Critical Roles for Distance, Task, and Motion in Space Perception: Initial Conceptual Framework and Practical Implications. Human Factors 50(5), 811–820 (2008)
12. DeLucia, P.R.: Pictorial and motion-based information for depth judgements. Journal of Experimental Psychology: Human Perception and Performance 17, 738–748 (1991)
13. Dornheim, M.A.: Dramatic Incidents Highlight Mode Problems in Cockpits. In: Aviation Week & Space Technology, January 30 (1995)
14. Forsythe, C., Bernard, M.L., Goldsmith, T.E.: Human Cognitive Models in Systems Design. Lawrence Erlbaum Associates, Mahwah (2005)
15. Freed, M.A., Remington, R.W.: Making human-machine system simulation a practical engineering tool: An apex overview. In: Proceedings of the International Conference on Cognitive Modelling (2000)

16. Gersh, J.R., McKneely, J.A., Remington, R.W.: Cognitive engineering: Understanding human interaction with complex systems. Johns Hopkins APL Technical Digest 26(4) (2005)
17. Gigerenzer, G., Todd, P.M.: The ABC Research Group: Simple heuristics that make us smart. Oxford University Press, New York (1999)
18. Gluck, K., Pew, R.: Modeling Human Behavior with Integrated Cognitive Architectures: Comparison, Evaluation, and Validation. Lawrence Erlbaum Associates, Mahwah (2005)
19. Gray, R., Regan, D.M.: Perceptual Processes Used by Drivers During Overtaking in a Driving Simulator. Human Factors 47(2), 394–417 (2005)
20. Hutchins, E.: Cognition in the Wild. MIT Press, Cambridge (1995)
21. Keeny, R.L., Raiffa, H.: Decisions with Multiple Objectives: Preferences and Value Trade-Offs. Wiley & Sons, New York (1976)
22. Klein, G.: Naturalistic Decision Making. Human Factors 50(3), 456–460 (2008)
23. Leiden, K., Laughery, K.R., Keller, J., French, J., Warwick, W., Wood, S.D.: A review of human performance models for the prediction of human error. Technical report, NASA, System-Wide Accident Prevention Program, Ames Research Center (2001)
24. Lüdtke, A., Möbus, C.: A cognitive pilot model to predict learned carelessness for system design. In: Pritchett, A., Jackson, A. (eds.) Proceedings of the International Conference on Human-Computer Interaction in Aeronautics (HCI-Aero), 29.09.-01.10 (2004)
25. Lüdtke, A., Osterloh, J.-P., Mioch, T., Rister, F., Looije, R.: Cognitive Modelling of Pilot Errors and Error Recovery in Flight Management Tasks. In: Palanque, P., Vanderdonckt, J., Winckler, M. (eds.) HESSD 2009. LNCS, vol. 5962, pp. 54–67. Springer, Heidelberg (2010)
26. Lüdtke, A., Osterloh, J.-P.: Simulating Perceptive Processes of Pilots to Support System Design. In: Gross, T., Gulliksen, J., Kotzé, P., Oestreicher, L., Palanque, P., Prates, R.O., Winckler, M. (eds.) INTERACT 2009, Part I. LNCS, vol. 5726, pp. 471–484. Springer, Heidelberg (2009)
27. Meyer, D.E., Kieras, D.E.: A computational theory of executive cognitive processes and multipletask performance: Part 1. basic mechanisms. Psychological Review 104, 3–65 (1997)
28. Newell, A., Simon, H.A.: GPS, a program that simulates human thought. In: Feigenbaum, E., Feldmann, J. (eds.) Computers and Thought (1961)
29. Newell, A.: Unified Theories of Cognition. Harvard University Press (1994); Reprint edition
30. Newell, A.: You can't play 20 questions with nature and win: projective comments on the papers of this symposium. In: Chase, W.G. (ed.) Visual Information Processing. Academic Press, New York (1973)
31. Norman, D.A.: Steps Toward a Cognitive Engineering: Design Rules Based on Analyses of Human Error. In: Nichols, J.A., Schneider, M.L. (eds.) Proceedings of the SIGCHI conference on Human factors in computing systems, pp. 378–382. ACM Press, New York (1982)
32. Nuxoll, A., Laird, J., James, M.: Comprehensive working memory activation in Soar. In: International Conference on Cognitive Modeling (2004)
33. Olson, J.R., Olson, G.M.: The Growth of Cognitive Modeling in Human-Computer Interaction since GOMS. In: Human-Computer Interaction, vol. 5, pp. 221–265. Lawrence Erlbaum Associates, Inc., Mahwah (1990)
34. Orasanu, J., Connolly, T.: The reinvention of decision making. In: Klein, G.A., Orasanu, J., Calderwood, R., Zsambok, C.E. (eds.) Decision Making in Action: Models and Methods, pp. 3–21. Ablex Publishing Corporation, NJ (1993)

35. Pew, R., Mavor, A.S.: Modeling Human and Organizational Behavior. National Academy Press, Washington (1998)
36. Ritter, F.E., Young, R.M.: Embodied models as simulated users: Introduction to this special issue on using cognitive models to improve interface design. International Journal of Human-Computer Studies 55, 1–14 (2001)
37. Salvucci, D.: A multitasking general executive for compound continuous tasks. Cognitive Science 29, 457–492 (2005)
38. Sarter, N.B., Woods, D.D.: How in the World did we get into that Mode? Mode error and awareness in supervisory control. Human Factors 37(1), 5–19 (1995)
39. Sarter, N.B., Woods, D.D.: Strong, Silent and Out of the Loop: Properties of Advanced (Cockpit) Automation and their Impact on Human-Automation Interaction, Cognitive Systems Engineering Laboratory Report, CSEL 95-TR-01, The Ohio State University, Columbus OH (1995)
40. Sarter, N.B., Woods, D.D., Billings, C.: Automation Surprises. In: Salvendy, G. (ed.) Handbook of Human Factors/Ergonomics, 2nd edn. Wiley, New York (1997)
41. Simon, H.A.: A Behavioral Model of Rational Choice. In: Models of Man, Social and Rational: Mathematical Essays on Rational Human Behavior in a Social Setting. Wiley, New York (1957)
42. Simon, H.A.: Administrative Behavior. A Study of Decision-Making Processes in Administrative Organization, 3rd edn. The Free Press, Collier Macmillan Publishers, London (1976)
43. Todd, P.M., Gigerenzer, G.: Putting Naturalistic Decisions Making into the Adaptive Toolbox. Journal of Behavioral Decision Making 14, 353–384 (2001)
44. Treat, J.R., Tumbas, N.S., McDonald, S.T., Shinar, D., Hume, R.D., Mayer, R.E., Stanisfer, R.L., Castillan, N.J.: Tri-level study of the causes of traffic accidents. Report No. DOT-HS-034-3-535-77, Indiana University (1977)
45. Treisman, A.M., Gelade, G.: A feature-integration theory of attention. Cognitive Psychology 12, 97–136 (1980)
46. Tversky, A., Kahneman, D.: Judgment under uncertainty: Heuristics and biases. Science 185, 1124–1131 (1974)
47. Wiener, E.L.: Human Factors of Advanced Technology ("Glass Cockpit") Transport Aircraft. NASA Contractor Report No 177528. Moffett Field, CA: NASA Ames Research Center (1989)
48. Woods, D.D., Roth, E.M.: Cognitive Engineering: Human Problem Solving with Tools. Human Factors 30(4), 415–430 (1988)
49. Woods, D.D., Johannesen, L.J., Cook, R.I., Sarter, N.B.: Behind Human Error: Cognitive Systems, Computers, and Hindsight. Wright Patterson Air Force Base, Dayton, OH: CSERIAC (1994)
50. Wray, R., Jones, R.: An introduction to Soar as an agent architecture. In: Sun, R. (ed.) Cognition and Multi-agent Interaction, pp. 53–78. Cambridge University Press, Cambridge (2005)
51. Zsambok, C.E., Beach, L.R., Klein, G.: A Literature Review of Analytical and Naturalistic Decision Making. Technical Report Klein Associates Inc., 582 E. Dayton-Yellow Springs Road, Fairborn, OH 45324-3987 (1992) `http://www.au.af.mil/au/awc/awcgate/navy/klein_natur_decision.pdf`

Integrating Collective Work Aspects in the Design Process: An Analysis Case Study of the Robotic Surgery Using Communication as a Sign of Fundamental Change

Anne-Sophie Nyssen and Adelaide Blavier

University of Liege, Laboratory of Cognitive Ergonomics,
5 boulevard du rectorat, B32, 4000 Liège, Belgium
asnyssen@ulg.ac.be

Abstract. Ergonomic criteria are receiving increasing attention from designers but their applications don't ensure that technology matches the system's constraints and its reliability. The aim of this paper is to study how robotic surgery induces fundamental changes in the collective work using communication as a sign of the adaptation processes. First, we compared verbal communication between surgeons in two conditions (laparoscopy and robotic surgery). Secondly, we compared three teams with different level of expertise with the robotic system on a repeated surgery act in order to identify permanent and transitory changes. Third, we analyzed conversion cases. We showed more acts of communication with the robotic system. The content analyses of communication revealed a profound change of the structure of the task that requires explicit collaborative modes. Although our sample is small, our results can be extended in other domains concerned with telework.

Keywords: Robotics, collective work, adaptation processes, design, assessment.

1 Introduction

The number, complexity and variety of medical devices have increased in recent years. At the same time, human error is considered to be the major contributing factor of medical accidents. Accident investigations are traditionally based on epidemiological methods rather than on detailed analyses of work situations. These methods often classify accidents into exclusive categories: human error, equipment failure or unavoidable complication. We can ask ourselves if such a classification still makes sense in our modern world where human, techniques and organization are interdependent. The health care system is characterized by diversity, complexity and the need for coordinated work between multiple disciplines. This has caused great difficulty in the design of clinical technical systems. Designers can be some kind of dreamers; they discover how difficult it is to assist activity in naturalistic situations. Many technical aids are not used, are misused or induce new forms of errors. This paradox was depicted by Bainbridge [1] for automated systems as the irony of automation. Among the reasons for these failures we can quote [2]: 1) a large mismatch between aid support and users' real needs, 2) the communication gap between potential users and computer science, for example, the role of the aid is often unclear for

The original version of this chapter was revised: The copyright line was incorrect. This has been corrected. The Erratum to this chapter is available at DOI: 10.1007/978-3-642-11750-3_10
P. Palanque, J. Vanderdonckt, and M. Winckler (Eds.): HESSD 2009, LNCS 5962, pp. 18–27, 2010.

the user, 3) the absence of a coherent design philosophy: for instance, the method of knowledge representation may be inappropriate, 4) the disregard of organizational issues: the complex environment where the system is used is not taken into account, nor are its dynamics and uncertainty.

Regarding the unintended side effects of technology, several researchers have indicated the need to reevaluate the human-machine interaction at a fundamental level [3, 4, 5, 6]. The concept of *user-centered design* refers to this attempt. The fundamental principles of such design approaches are: involvement of target-users in the design process, action-facilitation design and scenario-based design. Even if accepting the centrality of the user in the design process is becoming a more accepted prerequisite of appropriate person-machine design, its application has often been limited in practice to some particular design stages. A look at the design cycle schematized by Wickens, Gordon and Liu [7] illustrates the common practice of failing to involve the users. At the beginning of the cycle, potential users rarely converse with designers. It is the "human factors professionals", sometimes psychologists, sometimes ergonomists, who provide designers with the frame of reference concerning the task, the work environment and users' needs. As the prototype is developed, users are more easily included in the design process, especially for the validation of the prototype. At the end of the design process, the functionality of the product is assessed sometimes in real use, for a period of time. However, at this late stage, changing the product becomes unfeasible and procedures or training measures constitute, for the most part, the protective measures that ensure safety of the joint cognitive system.

Conducted in this way, none of the above stages relate specifically to a *user in context* centric view. The process places the product at the center.

From an activity theory perspective [8,9], aid systems should be designed to support operators in doing a task safely and efficiently in real work situations. Cognitive activity analysis as developed by Rasmussen and Vicente [10], is placed at the center of the analysis, focusing on information, mental effort, decision making and regulation. The concept of ecological interface was developed to illustrate an interface that provides appropriate support for the different levels of cognitive functioning.

Along the same line, but this time stressing the contextual and social point of view, is the *Scenario-Based Design approach*, a set of perspectives linked by a radical vision of user-oriented design [11]. This approach is not entirely new. For decades, systems developers have spontaneously used scenarios to envision future concrete use of their systems. But this informal practice has gained international acknowledgment, and the social content of the work is taken into account. To integrate context into the design, the task analysis stems from a scenario: "One key element in this perspective is the user-interaction scenario, a narrative description of what people do and experience as they try to make use of computer systems and applications. Computer systems and applications can and should be viewed as a transformation of user tasks and their supporting social practices" [11, pp 3]. Despite these valuable insights, scenarios constitute only examples of interactions of use and thus suffer from incompleteness.

We use one study to illustrate how important in-depth work analysis is in evaluating and designing new technology. More than 600 hours of observation were conducted in the operating rooms selected on the basis of their use of the new robotic system.

2 Robotic Surgery System

Surgery has known important developments with technological advances. Laparoscopy is certainly one of them. There is little question that laparoscopy represents a definite progress in patient's treatment. However, there are a lot of drawbacks, some of which are not without significance. For instance, the surgeon has lost all tactile feedback, (s)he has to perform operation with only sensory input from the two-dimensional picture on a video screen, and the procedure, to be done with long instruments, is seldom performed in a comfortable position for the surgeon. The fact that long instrument are used through an opening (trocar) in the abdominal wall, limits the degrees of freedom of the surgeon to a number of 4: in and out, rotation around the axis, up and down and from medial to lateral. The aim of the computer guided mechanical interface, commonly referred to as a robot, is to allow for 1) restoration of the degrees of freedom that were lost, thanks to an intra-abdominal articulation of the surgical tools, 2) three dimensional visualization of the operative field in the same direction as the working direction, 3) modulation of motion amplitude by stabilizing or by downscaling and 4) remote control surgery. Because of these improvements, the surgical tasks can be performed with greater accuracy.

However, to place a computer as an interface between the surgeon and the patient transforms the joint cognitive system.

Laparoscopy procedures typically involve the simultaneous use of three or more instruments (e.g. laparoscope, probe or gripper and shears or other cutting tools). Because of this, at least one tool must be operated by an assistant. The assistant's task is often limited to static functions of holding the instrument and managing the camera.

In classical laparoscopy, the assistant and the surgeon are face to face, and they use the same 2D representation of the surgical field to tailor the task.

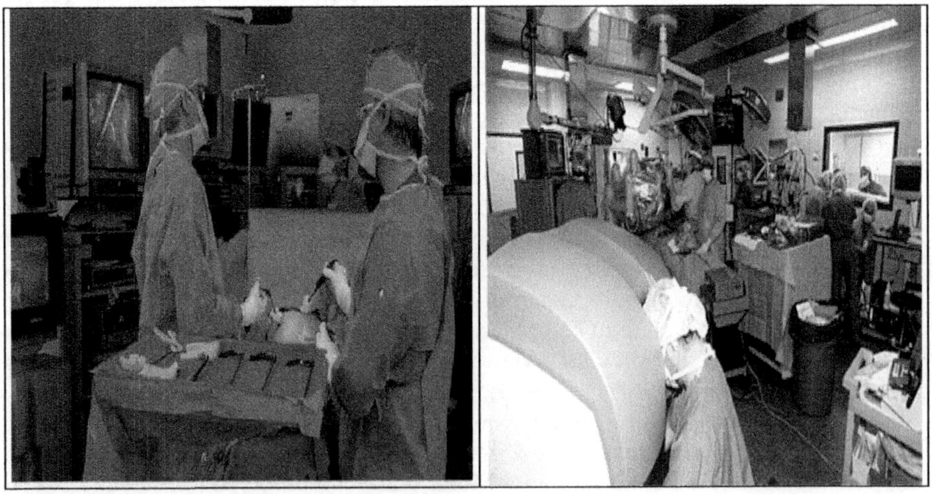

Fig. 1. Configuration of the operating theater in classical laparoscopy (left) and with the robotic system (right)

In robotic surgery, the surgeon is seated in front of the console at a distant point, looking at an enlarged three-dimensional binocular display on the surgical field while manipulating handles that transmit the electronic signals to the computer that transfer the exact same motions to the robotic arms. Robotic surgery can be performed at distant locations. However, within the actual technological system, the surgeon is still in the same operating room as the patient. The computer-generated electrical impulses are transmitted by a 10-meter long cable that controls the three articulated "robot" arms. Disposable laparoscopic articulated instruments are attached to the distal part of two of these arms. The third arm carries an endoscope with dual optical channels, one for each of the surgeon's eyes, which allows a true binocular depth perception (stereoscopy). The assistant is next to the patient, holding one or two instruments and looking at a 2-D display of the surgical field.

3 Communication as a Sign of Adaptation Requirements

Every act of communication, both verbal and non verbal, can be considered as an adaptive process analogous to biological evolution. Adaptation is the process of adjusting the mental structures and the behavior to cope with changes. Because so much of the adaptation processes in real time within the health care system are still verbal communication, the analysis of language becomes an important paradigm in order to study the adaptation capacities of a system facing a change.

When practitioners repeatedly work together, a reduction of verbal information exchanges is observed as practitioners get to know each other. Information taken directly from the work field replaces the verbal exchanges. Indeed, any regular action, parameter or alarm takes on the character of the "initiator" of verbal communication (12; 13; 14). Other studies (i.e. 15) have examined the relationship between communication and non routine situations in complex systems: the greater the trouble, the greater are the demands for information centered on the task across the members of the team.

Based on the above arguments, three important points can be noted. First, the environment provides feedback, which is the raw material for adaptation. Simple systems tend to have very straightforward feedback, where it often easy and instantaneous to see the result of an action. Complex systems may have less adequate feedback. The deployment of technology has increased the complexity of communication from non verbal to verbal, and to complex symbolic patterns. Additionally, introducing media and a distance between the agent and the process to control can delay and/or result in loosing feedback information. In laparoscopy surgery, the surgeon looses direct contact with the surgical site. S/he looses tactile feedback and performs operations with only sensory input from the video picture. As the robotic system is introduced in the OR, s/he looses proprioceptive feedback in addition to loosing a face to face feedback communication channel.

Secondly, communication is a dynamic feedback process which, in turn, affects the communicators. As we shall see, because the assistant and the surgeon have often prior knowledge and experience with the task, the assistant can anticipate the next movement or instrument that the surgeon needs in a routine task and non verbal communication can be very efficient (e.g., when the surgeon makes a hand signal to indicate to stop the movement or when s/he looks at the assistant to verify the receipt of an implicit request).

Third, in this dynamic perspective, short term adaptation feedback strategies that are exclusively based on verbal communication can be highly resource-consuming for the practitioners over time and, thus, may lead to long term inadequate adaptation.

Each of these points will be dealt with in our working hypotheses.

- In the case of adaptation, it is hypothesized that the technical system provides good feedback that supports the system to carry the task. Within our framework that views communication as an adaptive process, the following can be expected with the introduction of a robot system:
    - in the short term, new patterns of communication that reveal adaptation strategies
    - with training and regular interactions, a reduction of communication that reveals the dynamic nature of the adaptation process
- In the case of lack of or inappropriate adaptation, the technical system provides inadequate feedback resulting in increasing and maintaining the verbal communication to compensate for the weakness of feedback from the new equipment.

4 Experimental Study and Verbal Communication Analysis

We carried out three studies to examine our hypotheses:

1. First, we compared surgical operations that were performed with a robotic system compared with classical laparoscopy. We chose two types of surgery procedures (digestive and urology) because it is possible to perform them with either classical laparoscopy or with a robotic system.

In the two conditions (robotic and classical laparoscopy), the team members were identical. They were experts in the use of classical laparoscopy (>100 operations) and were at least familiar with the use of the robotic system (> 10 operations). We observed 5 cholecystectomy (digestive) with the robotic system and 4 with classical laparoscopy, and 7 prostatectomy (urology) with the robotic system and 4 with classical laparoscopy.
The robotic system used in our study was the Da Vinci robotic system (Intuitive Surgical, Mountain View, CE, USA) as shown in Figure1.

2. Secondly, we compared teams with different levels of expertise with the robotic system during gynecology surgery. We compared three teams with different levels of expertise who successively performed two tubular reanastomosis of 36 Fallopian tubes: 1) both the surgeon and the assistant were experts with the robotic system (>50 operations with the robotic system), 2) the surgeon was an expert while the assistant was a novice with the robotic system (<10 operations with the robotic system), 3) the surgeon and the assistant were novices with the robotic system (<10 operations with the robotic system).

3. Thirdly, we compared routine and non routine operations: conversion from robot surgery to classical surgery.

In the three studies, we recorded all the verbal communication between the surgeon and the assistant. We analyzed their content and identified six categories of communication. We also measured the duration of the intervention, as this is an important performance criterion for surgeons.

The six types of communication were:

- Verbal demands concerning the orientation (and localization) of organs.
- Verbal demands concerning the manipulation of instruments and/or organs.
- Explicit clarification concerning strategies, plans and procedures.
- Orders referring to tasks such as cutting, changing instruments, and cleaning the camera.
- Explicit confirmation of detection or action.
- Other communications referring to state of stress or relaxation.

For each category, we measured the number of acts of communication, while taking into account the duration of the surgery (ratio = number of acts of communication / time (in seconds) X 100). The Mann-Whitney *U* test was used to compare the two techniques: classical laparoscopy and robotic surgery and the Kruskal-Wallis test was used across the board.

5 Results

5.1 Communication as a Feedback Adaptive Process

The average duration of the intervention was significantly longer ($p<0.05$) with the robotic system (cholecystectomy: 82.59±27.37; prostatectomy: 221.39±58.79) than with classical laparoscopic (cholecystectomy: 31.85±9.64; prostatectomy: 95.74±11.53).

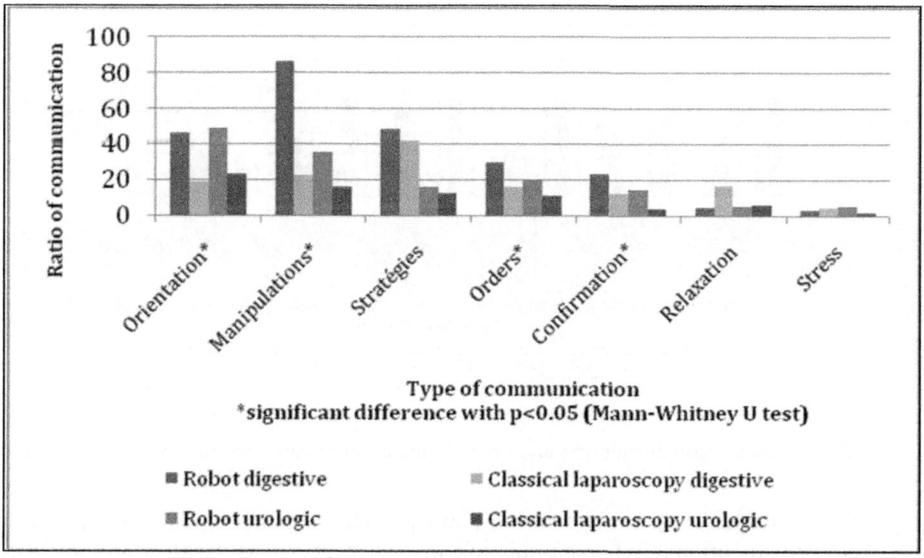

Fig. 2. Communications during robotic and classical laparoscopy in digestive and urologic surgery

Figure 2 shows that the introduction of the robotic system created a new pattern of communication. This pattern of communication was similar for the two types of surgery.

The significant increase in the number of communication acts ($p<0.05$) referring to orientation, manipulation, order and confirmation within the robot system suggests that a breakdown occurs in the collaboration between the surgeon and the assistant. The surgeon works alone and continually needs to ask the assistant about the orientation and the placement of the instrument (which is manipulated by the assistant) in order to facilitate the identification of the organs as demonstrated in the following example of interaction:

Surgeon at the consol: "could you tell me if you are touching something here, because I see a particularity "

Assistant surgeon near the patient: "yes, I am touching something hard - it is a bone".

Explicit demands, order, and confirmation are needed because the system configuration impedes face to face implicit control and anticipation of the actions.

5.2 Communication as a Dynamic Adaptation Processes: Permanent and Transitory Changes

Our experimental plan allows us to identify the permanent and the transitory changes induced by the change of equipment.

Our results show that the number of acts of communication is reduced with repeated experience: from the first operation to the second operation of Fallopian tube anastomosis, but also with the degree of expertise of the team with the robotic system (see fig. 3).

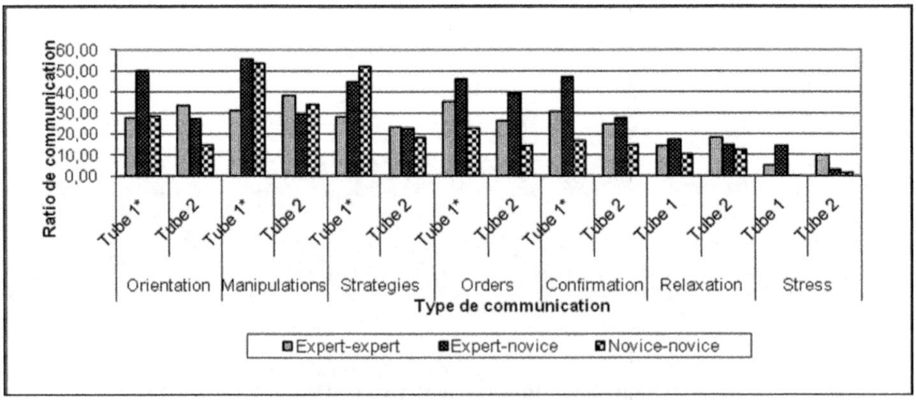

Fig. 3. Communication during first and second tube anastomosis according to the expertise

Detailed analysis of communication showed that the number of communication acts referring to orientation, manipulation and strategies was significantly reduced ($p<0.05$) when both surgeons were experts in robotic surgery and from the 1st tube to the 2nd tube. Not surprisingly, the number of acts of communication referring to order and confirmation was significantly greater when an expert was present in the team.

We observed that this increase of order and confirmation does not change from the 1st tube to the 2nd tube and is maintained within the experts' team.

5.3 Communication as a Sign of Trouble

We observed two conversions: 1 in urology from a robotic surgery to open surgery and 1 in digestive surgery from robotic surgery to classical laparoscopy surgery.

Each of these conversions is associated with an increased number of verbal communications (see fig.4). These communications concerned explicit clarification of strategies (replanning) and expectations concerning orientation and manipulations. We also observed less communication that referred to confirmation. During a crisis, the surgeon seems to act; he does not take the time to verify the receipt of his action or request.

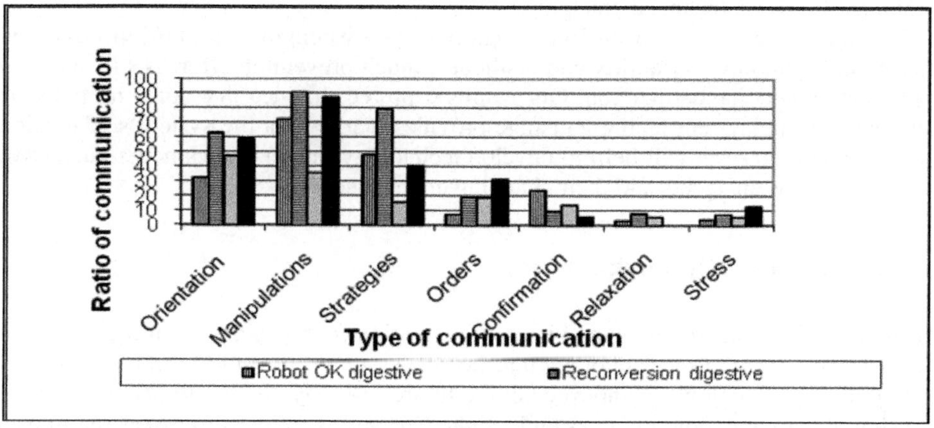

Fig. 4. Communication during first and second tube anastomosis according to the expertise

6 Discussion

Based on our results, it is clear that the robotic system changes the feedback loop and that verbal communication used by surgeons is a feedback-adaptive process to compensate the feedback information absent in the robotic system. Our results show that both the number of communication acts and the type of communication evolves with the agent-robot environment interactions, suggesting some kind of successful adaptation to the change of equipment. It seems that manipulation, orientation and strategies can be rapidly learned through interaction with the technical system (from tube 1 to tube 2.). However, orders and confirmation are maintained within the experts' team. This result suggests that, by introducing a distance between the surgeon and the assistant, the robotic system changes profoundly the structure of the task and the mode of cooperation between the surgeon and his assistant. It favors an explicit division of work and an explicit leadership based on order and continual control of the work (confirmation). As a result, the status of the assistant and of the OR team are modified. The surgeon assistant becomes more like a technician, responding to the orders of the surgeon. There are two new actors in the team: the robot and the robot technician who become essentials. We can predict an impact of these changes on the OR team work satisfaction associated with new forms of errors such as a loss of "situational awareness".

As mentioned earlier, when complications occur, increased verbal communication is required to clarify plans and expectations in order to enable rapid coordinated actions between the surgeon and the assistant and to maintain a update shared situational awareness. These conversion cases show how the surgeons, and not the robot, have mechanisms for recovering from the situation before it affects the patient by replanning the cases into classical surgery. This means that the system's capacity for facing unexpected events resides in the human part rather than in the technical part of the system. Indeed, adaptation emerges through the history of different agent-environment coupling over time (open surgery, classical laparoscopic surgery, robotic surgery) that enhances the agent's autonomy towards the variability from the environment (for eg. a technical change).

Although recent work from Joint Cognitive systems engineering [16] discusses issues like autonomy, variability and resilience, much prevention effort is still spent on automation and standardization. Our results captured the idea that studying both the behavior of the system facing a change provides markers on the system's adaptation capacity and, in turn, will help to develop technology that enhances better adaptative coupling between agents and their changing environment.

7 Concluding Remarks

Ergonomic criteria are receiving increasing attention from designers but their application doesn't ensure that technology matches the system's constraints and enhances its reliability. Although we cannot predict the future, we may attempt to better guide the design process by adopting a systemic view. Our aim is to insist to assess the impact of technology changes on all the dimensions of a work situation: technical, economic, performance, cognitive, and organizational. In the health care system, as in other complex and dynamic systems, there is a need for researchers and designers to more investigate the impact of the equipment on the reciprocal interaction between cognition and organization. Doing so is critical for the quality, safety and effectiveness of the modern work.

References

1. Bainbridge, L.: The ironies of automation. In: Rasmussen, J., Duncan, K., Leplat, J. (eds.) New technology and human error, pp. 271–283. Wiley, London (1987)
2. O'Moore, R.: The conception of a medical computer system. In: van Gennip, E.M.S.J., Talmon, J.L. (eds.) Assessment and evaluation of information technologies in medicine, pp. 45–49. IOS Press, Amsterdam (1985)
3. Sheridan, T.B.: Human centered automation: Oxymoron or common sense? presented at the industrial summer school on human-centered automation, Saint-Lary, France (1995)
4. Norman, D.A.: The invisible computer. MIT Press, Cambridge (1999)
5. Norman, D.A.: Things that make us smart: Defending human attributes in the age of the machine. Addison-Wesley, Reading (1993)

6. Billings, C.E.: Aviation automation: The search of a human-centered approach. Lawrence Erlbaum Associates, Mahwah (1997)
7. Wickens, C.D., Gordon, S.E., Liu, Y.: An introduction to human factors engineering. Longman, New York (1997)
8. Roe, R.A.: Acting systems-design – an action theoretical approach to the design of man-computer systems. In: De Keyser, V., Qvale, T., Wilpert, B., Ruiz Quintanilla, S.A. (eds.) The meaning of work and technical options, pp. 179–195. Wiley, Chichester (1988)
9. Hacker, W.: Activity: A fruitful concept in industrial psychology. In: Frese, M., Sabini, J. (eds.) Goal directed behavior: The concept of action in psychology, pp. 262–383. Lawrence Erlbaum Associates, Hillsdale (1985)
10. Rasmussen, J., Vicente, K.L.: Coping with human errors through system design: implications for ecological interface design. International Journal of Man-Machine Studies 31, 517–534 (1989)
11. Caroll, J.M.: Scenario-based design. Envisioning work and technology in system development. Wiley, New York (1997)
12. Savoyant, A., Leplat, J.: Statut et fonction des communications dans l'activité des équipes de travail (Statut and function of the communications in the activities of the workteams). Psychol.fr. 28(3), 247–253 (1983)
13. Pavard, B.: Système coopératifs: de la modélisation à la coopération. Octares, Toulouse (1994)
14. Nyssen, A.S., Javaux, D.: Analysis of synchronization constraints and associated errors in collective work environments. Ergonomics 39, 1249–1264 (1996)
15. Bressolle, M.C., Decortis, F., Pavard, B., Salembier, P.: Traitement cognitif et organisationnel des micro-incidents dans le domaine du contrôle aérien: analyse des boucles de régulation formelles et informelles. In: De Terssac, G., Friedberg, E. (eds.) Coopération et conception, Octares, Toulouse, pp. 267–288 (1996)
16. Woods, D.D., Hollnagel, E.: Joint Cognitive Systems: Patterns in cognitive systems engineering. Taylor & Francis Group, Boca Raton (2006)

Patient Reactions to Staff Apology after Adverse Event and Changes of Their Views in Four Year Interval

Kenji Itoh[1] and Henning Boje Andersen[2]

[1] Dept. of IE and Management, Tokyo Institute of Technology
2-12-1 Oh-okayama Meguro-ku, Tokyo 152-8552, Japan
Itoh.k.aa@m.titech.ac.jp
[2] Dept. of Management Engineering, Technical University of Denmark
Produktionstorvet 426A, 2800 Kgs. Lyngby, Denmark
boje@man.dtu.dk

Abstract. In the present paper we report results of a patient survey about safety related issues carried out in 2007 in Japan focusing on patient attitudes to receiving different kinds of apology from healthcare staff after a medical accident. Results show, first, that the strongest preference of patients is for a "full" apology including a hospital promise of taking preventive actions against similar events in the future; and second, that the least effective reaction by healthcare staff is a so-called "partial" apology in which staff express sympathy or regret about the event, and which is in fact perceived as worse than "no apology", i.e., merely informing the patient about the event and future health risk. Comparing results to a similar survey in 2003, it appears that since then Japanese patients' perceptions of healthcare professionals and organisations, though still not very trustful, have changed slightly to a more positive point of views.

Keywords: Patient views; Apology; Adverse event; Patient questionnaire.

1 Introduction

There has been a massive increase in attention to patient safety in Japan in recent years. One of the signs of the attention has been the extensive coverage in the Japanese press of often tragic and sometimes spectacular instances of "medical errors". For instance, the major Japanese newspapers brought stories about 412 medical accidents in 2001 [1]. Since then, the number of media reports in newspapers and broadcasts has continued to increase. According to another survey [2], a total of 655 accident cases were reported by one of the largest Japanese news agencies in the entire year of 2005. Among these only 16% were cases that occurred in the same year, and approximately 60% concerned events that had happened more than three years earlier. Such repeated press reports about both recent accidents and, not least, accidents that have happened long ago indicate increasing public concerns with patient safety issues. At the same time, the Japanese public receives from the press an impression that hospitals often deal defensively, passively or inadequately when

The original version of this chapter was revised: The copyright line was incorrect. This has been corrected. The Erratum to this chapter is available at DOI: 10.1007/978-3-642-11750-3_10

P. Palanque, J. Vanderdonckt, and M. Winckler (Eds.): HESSD 2009, LNCS 5962, pp. 28–43, 2010.

patients are injured due to medical error, and the public gets the impression that they are slow and reluctant to reveal facts and apologise to the injured patients and their families. The picture seems to be similar to that of some other countries (e.g., [3], [4]).

Hospitals and hospital staff do not have an evidence-based approach to dealing with reactions to patients after adverse events. Consequently, when Japanese hospitals make plans for dealing with adverse events, efforts tend to be directed towards litigation defences and not preparing staff to deal openly and directly with patients and their families. Although a number of studies, but few Japanese, have been made of patients' views and requirements following an adverse event (e.g., [5]—[8]) little is known about patient priorities. Moreover, several studies show that differences in national culture make it questionable to transfer results across cultural borders (e.g., [9], [10]).

Accordingly, in a previous study [11], we investigated Japanese patients' views and recognition of safety related issues, in particular their expectations of disclosing actions taken by healthcare staff when suffering medical errors, compared not only with the original Danish survey [12], but also with a survey about the same actions taken by Japanese doctors [13]. Results of these studies indicated that Japanese patients were suspicious about the willingness of healthcare staff to disclose adverse events. Two potential reasons were also suggested as major potential sources behind the Japanese patient's mistrusts, namely, the steady stream of uniformly negative media reports and actual staff reluctance against openness [11]. It was also suggested that the public's views about patient safety issues were influenced to some extent by media reports as well as people's own experience healthcare contact [11].

During recent years, Japanese hospital managers have undertaken strong initiatives to enhance patient safety and include also encouraging hospital staff to deal with patients after medical mistakes and accidents. At the same time, a slight change has happened in the style and contents of media reports about healthcare issues, although there have still been extensive press reports of medical accidents. Recently, press reports have started describing situations in hospitals that readers will regard as appalling and extremely bad work conditions for healthcare professionals: lack of healthcare professionals – particularly hospital doctors – and therefore long working hours and hardly any days off. So, readers are likely to appreciate that there is an extraordinarily high workload in healthcare settings. Based on the above, one may therefore expect that patients' views may have become perhaps somewhat more positive since over the last five or so years; and more specifically, that patient expectations to staff actions after an adverse event will have become more favourable and that patients will generally show greater appreciation of efforts by staff and management to control safety.

Given the background as just outlined, it is of interest to investigate changes in patient views of and attitudes to healthcare staff and management over recent years. We particularly need to explore patients' willingness to forgive mistakes by healthcare staff and organisations [14], [15], e.g., investigate the degree of importance that patients assign to receiving an apology after an injury caused by medical error among different types of staff reactions. We therefore conducted a questionnaire-based survey in which we collected approximately 1,750 responses from patients and families in 14 Japanese hospitals. The survey data were also compared with those of the former survey [11] to analyse changes in patient views in the last four year interval.

In the present paper we report the results of the new survey, focusing on patients' attitudes to receiving staff apology after a medical accident. Changes in patient views of safety related issues in the last half decade are also discussed based on a comparison of the 2003 and the 2007 survey. Finally, some possible ways are also discussed to improve patient views of healthcare professionals by pursuing a patient-centred approach to risk management in healthcare.

2 Questionnaires and Respondents

The questionnaire comprised seven sections, of which the present paper focuses on just the following two aspects of patient views: (1) expectations about a doctor's reactions after an adverse event, and (2) the likelihood of the respondent acceptance to the doctor's apology. The present paper also describes briefly responses to two other sections and compares them to those in the previous study [11]: perceived causes of medical errors, and patient views and recognition of safety related issues in healthcare. An additional demographic section asked respondents to supply information about clinical department, gender, age group, and recent experience of hospitalisation and whether they had suffered medical errors.

In the two main sections of the questionnaire respondents' reactions were elicited as responses to two fictitious adverse events (vignettes) – one in which the patient suffers a relatively severe outcome and the other a relatively mild outcome. The two fictitious cases, originally designed for a survey of staff attitudes to error reporting [12], [16], were stated as follows:

Case A (Mild outcome): A patient is hospitalised for planned elective surgery. Before his operation the patient will as a matter of routine for an elder or middle-aged patient receive an anticoagulant injection as a prophylactic against thrombosis. When dictating to the case notes, the doctor is interrupted several times due to patients suddenly getting ill, and the doctor forgets to include the anticoagulant for the patient. The patient develops a thrombosis in a vein in his left leg. He therefore has to remain hospitalised an additional week. It is very unlikely that he will have permanent impairment from the thrombosis.

Case B (Severe outcome): A patient is hospitalised in order to receive chemotherapy. The drug has to be given as a continuous infusion intravenously. There is no pre-mixed infusion available in the department and the doctor has to prepare it himself. While he is preparing the infusion, he is distracted. By mistake he prepares an infusion with a concentration 10 times greater than the prescribed level. The doctor does not discover the error until he administers the same drug to another patient later that day. By this time the patient has already received all of the high concentration infusion. He is aware that in the long term the drug may impair cardiac functioning. He realizes that there is a significant risk that the patient's level of functioning will be diminished and that she probably won't be able to maintain her present work.

In the first section of the questionnaire, each of the cases was followed by questions asking respondents to what extent they would expect a doctor to carry out each of the following potential actions: (1) keep it to himself/herself that he/she has made a mistake; (2) write in patient's case-record about the event; (3) inform the patient about the adverse event and the future risk; (4) explain to the patient that the event was caused by his/her mistake; and (5) apologise about the event to the patient. Response options were given on a five point Likert-type scale, ranging from 'definitely yes' to 'definitely not'.

The questions pertaining to the two cases also asked respondents about their likelihood of receiving the doctor's *apology* (in Section 2). The context in which respondents react to any such question should be briefly outlined: In Japan patients may consult or be treated at any hospital or clinic they wish, i.e., there is free access to any hospital. Almost all Japanese citizens join public health insurance, and patients will typically pay themselves 30% of the expenses of hospital treatment, while the remainder is covered by insurance or public taxpayer funding. Therefore, we assume if a person who was a victim of an adverse event will be more likely to return to the hospital or clinic in question when needed, if the person receives a reaction (e.g., a properly expressed apology) from the doctor who was involved the case. Thus, respondents were asked to rate their likelihood of the statement on a five point Likert type scale, "I will come back to this hospital for consultation next time if the doctor undertakes this reaction" shown in Table 1. For brevity, we will refer to the six reactions listed under Apology statements as just "apologies".

Table 1. Different possible reactions by doctor in vignettes rated by respondents

	Statements in the questionnaire	Abbreviation
(a)	Explain about the event that you have suffered and its consequence (and in Case B, also future risks)	Event explanation
(b)	Express sympathy to you about the event	Express sympathy
(c)	Express sympathy and apologise to you, admitting that the hospital must take responsibility for the event	Express apology
(d)	Offer exemption of expenses for additional treatment after the event, but no apology to you	Offer of fee exemption
(e)	Express sympathy and apologise to you, and offer exemption of expenses for additional treatment after the event	Express apology + Offer of fee exemption
(f)	Express sympathy and apologise to you, and promise you that the hospital will take action to avoid repetition of the incident	Express apology + Take preventive action

The survey was made in March-September of 2007 in Japan. We collected a total of 1,744 responses (overall response 46%) from inpatients and outpatients as well as families and relatives from 14 hospitals in Japan. Among these hospitals, nine hospitals are private and belong to the same owner group, and were located in Tokyo or near Tokyo area. The other five hospitals are public, belonging to local municipalities in non-metropolitan regions (but one of them located in the Chiba prefecture next to Tokyo). A paper questionnaire was given to each respondent by hospital administration staff, and filled-out questionnaires were returned by post (in pre-stamped envelopes) by the patient or relatives. The survey was anonymous, as was stressed to respondents.

3 Patient Attitudes to Staff Apology

3.1 Overall Attitudes of Japanese Patients

Respondents' answers to the question about the likelihood of their coming back to the same hospital when needed next time are summarised in Figure 1, which shows the proportion of respondents who accept or reject to each type of the doctor's apology. The bars represent percentages of both respondents who "definitely" or "probably" would return to the hospital to the right of the vertical dividing line ("accept") and those who "definitely" or "probably" would not return on the left side ("reject") for each of the specified apology actions. Statistical tests between the two cases (Mann-Whitney test) show that a significantly greater proportion of patients reject the doctor's apology when they suffered the severe outcome (Case B) than the mild outcome case (A), regardless types of apology taken by the doctor ($p<0.001$).

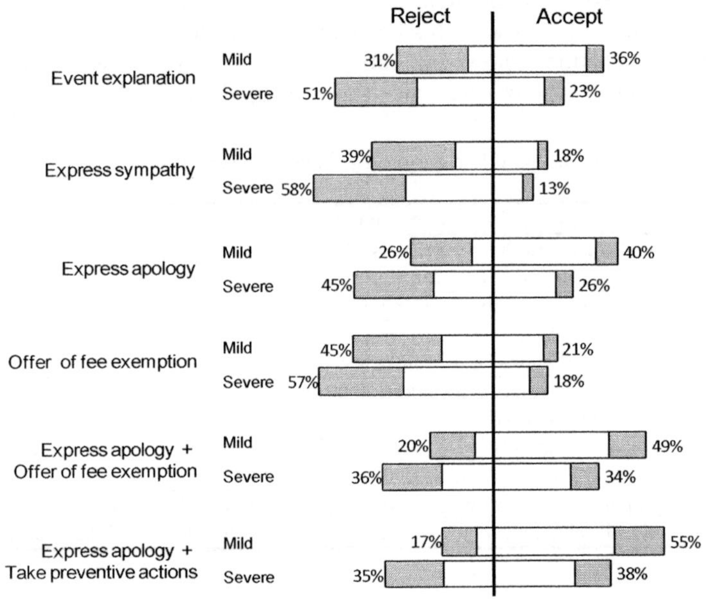

Fig. 1. Percentage of patient acceptance of each type of doctor apology

There was also a highly significant difference in patient acceptance or rejection of the staff apology between reaction types in each of different severity cases ($p<0.001$). The most "effective" apology reaction in terms of patient forgiveness was to express apology including admittance of hospital responsibility for the event plus a promise of taking actions for preventing against similar events in the future. Further, 55% and 38% of respondents stated their willingness to definitely or probably have their next appointment at the same hospital for the mild and the severe case, respectively. In contrast, the least effective reaction was merely to express sympathy to the patient, i.e., giving a merely "partial apology" among the six kinds of apology reactions offered to respondents – including no apology but explanation of the event: only 18% and 13% of respondents indicated they would return to the hospital for the minor and the severe outcome case, respectively. A large increase in percentage acceptance was obtained when expressing apology which includes admittance of the hospital responsibility to the adverse event, compared with expressing sympathy: a rise from 18% to 40% of acceptance for the mild (p<0.001), and 13% to 26% for the severe outcome case (p<0.001).

Similar to "expressing sympathy", the offer of fee exemption produced a very modest proportion of acceptance and a high proportion of rejection. There were no significant differences in percentage acceptance or rejection between these two reactions (expressing sympathy and fee exemption) for the mild (p=0.070) and the severe case (p=0.215). However, when the hospital staff expressed apology, the offer of exemption of additional expenses became much more effective to get the patient to accept the apology reaction, increasing acceptance from 40% to 49% for the mild (p<0.001) and from 26% to 34% for the severe case (p<0.001). Compared with only offering the exemption of additional expenses, the percentage acceptance was approximately doubled, and rejection decreased greatly in both cases. However, comparing "expressing apology" with either fee exemption or a promise of preventive measures showed little difference for the severe case (p=0.074), although there was a statistically significant preference for the latter in the mild case (p<0.001).

3.2 Differences by Patient Attributes

The degree of acceptance of each kind of doctor apology is provided across patient attributes in Table 2, divided into: types of medical errors respondents have experienced (major, minor vs. no error), age groups (below 60 vs. 60 and above), gender, and hospitalisation history within the last two years. Regarding the error types experienced, merging major and minor errors into a single "error experienced" group, responses of this group were significant different with those of the "no error experienced" group for some types of apology. The "error experienced" group exhibited more negative or less favourable attitudes to apology acceptance than the "no error experienced" group. However, for each of the apology types, no significant difference was identified between the "major error" and the "no error experienced" group. In contrast, percentage acceptance of the "major error" group was higher to any type of apology than that of the "minor error" group, although we observed significant differences between these two groups for only a few apology types – possibly due to small sample size of the "error experienced" respondents. One possible reason for more negative attitudes of the "minor error" than that of the "major error" group might possibly – and as a matter of speculation – be sought along the following line: Based on the descriptions of the "minor error" patients themselves, about half of whom described the error they had experienced, the errors are mostly near-misses or incidents with no effect on the patient, and some might perhaps be regarded as causes of nuisance though still perceived as an "error" by the patient. One may therefore expect that these patients are less likely to have been informed about errors by the doctor than patients that have experienced "major error" – most of these respondents provided detailed descriptions about their incident cases. Therefore, one may conjecture that respondents of the "minor error" group might become more sceptical against reactions of healthcare staff and organisation connecting after an adverse event.

Dividing respondents into two groups by their age (below 60 and 60 and above, which is the formal retirement age in Japan), it was seen that the younger group had slightly but significantly more negative attitudes to most types of the doctor apology than the older. This age trend was also seen for patient expectations of the doctor's disclosure actions [11] and those about the quality of care in healthcare [17]. This modest link between age and attitudes was also statistically significant for many types of apology, in particular for every type for the severe outcome case, when we did the rank-based Kruskal-Wallis test between seven age classes grouped in ten-year intervals.

Table 2. Percentage of patient acceptance to each kind of doctor reactions

Patient attributes	Express sympathy		Offer of fee exemption		Express apology		Exp. apology+ Take prev. act.	
Case	A	B	A	B	A	B	A	B
Suffered major error	26%	9%	22%	22%	43%	27%	53%	43%
Suffered minor error	9%	5%	19%	8%	32%	14%	43%	21%
No error experienced	18%	14%	23%	20%	43%	28%	59%	42%
$p^{\#}$ (Mann-Whitney)	**	*			*		*	*
< 60 years old	17%	10%	19%	14%	41%	21%	56%	32%
≥ 60 years old	19%	18%	23%	23%	39%	33%	54%	46%
p (Mann-Whitney)	**	***	**	***		***		***
Male	19%	15%	25%	20%	43%	30%	55%	43%
Female	16%	11%	18%	15%	37%	22%	54%	34%
p (Mann-Whitney)		**	***	***	**	***		***
Experience of hospitalisation last 2 yr.	18%	14%	22%	19%	42%	27%	57%	40%
No experience of hospitalisation last 2 yr.	16%	12%	17%	15%	36%	23%	51%	34%
p (Mann-Whitney)	*		*		**		**	*

#: between respondents suffering major or minor error and those having no error experience.
*: $p<0.05$, **: $p<0.01$, ***: $p<0.001$.

Regarding gender, significant differences were identified for most types of apology for the two cases of different severity. As a whole, female respondents were less liable to accept any of the apologies than male. This gender difference in patient attitude to medical staff reaction is directly opposite to the trend we have found in patient expectations to a doctor's disclosure actions: female patients exhibited more positive expectation of the doctor's error reporting actions and interaction with patients than male patients [11] – a finding that is matched in the present study, as will be described in the next section. For the other respondent attributes we obtained matching trends between patient expectation of the doctor's actions and their attitudes to receiving the doctor apology.

The questionnaire also asked respondents to state their hospitalisation experience within the last two years. The results shown in Table 2 suggest that recent experience of hospitalisation contributes to patient's positive attitudes to receiving doctor apology, in particular when an outcome of an adverse event is not severe. For the severe outcome case, a significant difference in the patient attitudes was observed between these two hospitalisation groups only when the doctor expressed an apology with a promise of preventive organisational actions.

We found no significant difference between clinical specialties that patients had consulted for any apology type, in contrast to the survey about patient expectations of staff actions after the adverse event [11].

4 Changes in Patient Views over Four Years

4.1 Expectations of Healthcare Staff Actions

To ascertain possible changes in patient views of safety issues over the last four-year interval we performed comparative analyses, using a dataset collected for the previous survey [11], in which we applied a similar questionnaire that shared the following sections with the one used in the present study: expectations about doctor's actions after an adverse event; causes of medical errors; and patient-safety related issues. The previous survey was conducted in 2003, collecting approximately 900 responses (64% overall response rate) from inpatients and outpatients in a university hospital in Tokyo.

Table 3. Changes in patient views about doctor's action after adverse event (2003/2007)

	Mild (A)			Severe (B)		
Doctor's potential actions	2007	2003	*p*	2007	2003	*p*
Keep it to him/herself	27%	32%	*	27%	27%	
	46%	43%		51%	47%	
Write in patient case record	39%	33%	**	38%	34%	
	36%	42%		34%	37%	
Inform patient of event & risks	48%	44%	**	47%	44%	
	30%	37%		28%	31%	
Admit own error	39%	32%	***	43%	36%	
	38%	45%		33%	36%	
Apology to patient	43%	34%	***	48%	41%	*
	33%	43%		27%	29%	

Upper row: % agreement (% of respondents stating 'definitely yes' or 'yes, probably').
Lower row: % disagreement (% of respondents stating 'definitely not' or 'probably not').
*: $p<0.05$, **: $p<0.01$, ***: $p<0.001$.

Comparative results of patient expectations about the doctor's actions are summarised in Table 3 in terms of percentage agreement and disagreement of each action item in 2007 and 2003 as well as results of the Mann-Whitney test between these two survey samples. As can be seen from this table, patient expectations about a doctor's reactions have been significantly improved for the mild outcome case (A) within the last four years. Compared with the 2003 responses, a smaller percentage of patients, on the one hand, expected the doctor involved in the event to keep it to himself or herself, but a greater part, on the other, took more favourable views of the staff actions such as writing the event into patient's case record, informing the patient about the event and its consequence, admitting one's own error for the event, and express one's apology to the patient in 2007.

For the severe outcome case, no significant difference was observed between the 2003 and the 2007 sample except for the doctor's apology to the patient. A greater percentage of patients agreed definitely or slightly that the doctor would apologise to the patient. Summarising the results mentioned so far, there is an overall trend that patient suspicion of healthcare staff action has been slightly reduced over the last four years.

The above results show that patient expectation to healthcare staff actions has become more favourable or trustful of staff during the last half decade.

4.2 Patient Safety Related Issues

The patient responses to eight safety-related questions at each of the two different survey periods, 2003 and 2007, are summarised in Table 4. Significant differences were observed in responses to many items on safety issues between 2003 and 2007. Among the items to which there was no significant change between the two surveys, two items were issues related to hospital reactions to a patient having suffered an adverse event, showing extremely high ("a right to be informed") and very high ("should receive compensation") percentage agreement. This might suggest that the Japanese population wish to have such requirements as a universal condition for healthcare systems independent of the present manner in which safety and quality are managed. In addition to these requirements about actions after adverse events, there was another requirement that was very high: the skills and competence of hospital staff should be tested regularly. However, the agreement with this statement has been highly significantly decreased from 2003. This may indicate that public trust in the competence of healthcare professionals has become stronger. As well as the extensive press coverage of patient safety problems and adverse events, there has been a marked increase of press reports about the lack of hospital professionals and their hard work in Japan, as mentioned in Section 1. These reports have particularly mentioned clinical specialties such as obstetricians, paediatricians, emergency doctors and anaesthesiologists, but also described the lack of staff in general. Such press reports as well as professional and organisational efforts in patient safety may possibly induce changes in patient views, changing them to a more positive perception of the competencies of healthcare professionals. But, as we now shall see, there has also been a change to more positive perceptions of other aspects.

Among the particularly salient patient responses to the safety-related issues is patients' high agreement with the statement "anyone can make a mistake". No significant difference was identified in this item between 2003 and 2007 – about 70% agreement and 10% disagreement. Referring to these responses, we suggest that majority of patients have realistic views of and recognition of human fallibility and (at least minor) errors that occur in the healthcare setting. This realistic patient recognition of human errors chimes in with the low agreement with the statement that "a ward/department that reports few errors can be expected as well to have few errors", although percentage agreement with this item increased slightly from 2003 to 2007.

Table 4. Changes in patient views about patient safety related issues (2003/2007)

Items	2007	2003	*p*
Staff skills & competence should be tested regularly	73% 11%	83% 6%	***
Patients have a right to be informed when an adverse event occurs	95% 2%	95% 3%	
Anyone can make mistake	71% 11%	72% 10%	
The press deals with medical errors in sensationalist way	55% 16%	44% 23%	***
Doctors cover up for each other	43% 15%	58% 8%	***
Few error reports can be expected as well making few errors	17% 36%	13% 37%	*
Individual staff committing error feels miserable about it	73% 7%	65% 9%	**
Patients suffering injury should automatically receive compensation	80% 4%	80% 4%	

Upper row: % agreement (% of 'definitely yes' or 'yes, probably').
Lower row: % disagreement (% of 'definitely not' or 'probably not').
*: $p<0.05$, **: $p<0.01$, ***: $p<0.001$.

Besides the above-mentioned items, there were a larger number of respondents who agreed than disagreed with the statement that the press generally deals with medical errors "in a sensationalist way". In addition, slightly less than three quarters of respondents (in 2007) showed their sympathy with healthcare staff involved in an adverse event, agreeing with the statement "the individual doctor or nurse who commits an error feels miserable about it". There were significantly large increases of percentage agreements for these issues from 2003 to 2007. In contrast, there was a significantly large drop in percentage agreement with the item that "doctors cover up for each other" during the four year period, which, again, suggests that patient views have become favourable to healthcare professionals. Based on these questionnaire results, we therefore suggest that Japanese patients seem to have a reasonable awareness of the human element involved in patient safety issues, and that this awareness has been somewhat strengthened during the last half decade.

4.3 Perceived Error Causes

Using the 2003 survey data, three causal factors on medical errors were elicited by the principal component analysis: staff workload, staff ability, and lack of management efforts [18]. Applying the same method to the 2007 data, we identified exactly the same construct of causal factors with 64% of cumulative variance accounted for. In this subsection, we mention comparative results of perceived causes of medical errors between 2003 and 2007 for nine individual items. Patient responses to each individual error causes are shown in Table 5 in terms of percentage agreement and disagreement as well as results of the Mann-Whitney test between 2003 and 2007. As can be seen in this table, there were significant differences in responses to all individual error causes between the two survey periods.

In terms of strength or weakness of each individual error cause, an overall trend has not been greatly changed. Individual items within the top five ranks as error causes were shared between the surveys in 2003 and 2007, i.e., "great workload", "fewer nurses", "fewer doctors", "inexperienced staff left" and "managements do little for safety". In addition, the top four items were forceful error causes – with which more than 50% of respondents agreed – in 2007, and the 2003 survey shared three items of these individual causes as forceful (cf. Table 5).

Table 5. Changes in causes of medical accidents perceived by patients (2003/2007)

Items	2007	2003	*p*
Working under great workload	76% 12%	71% 14%	**
Fewer nurses than really required	74% 11%	65% 13%	***
Fewer doctors than really required	71% 11%	47% 19%	***
Staff not sufficiently responsible for tasks	37% 34%	44% 28%	***
Staff is not sufficiently competent	28% 38%	30% 33%	*
Inexperienced staff is often left with insufficient back-up	61% 13%	66% 10%	***
Bad doctors are allowed to continue working	36% 27%	41% 24%	*
Hospital managements do little to prevent errors	37% 26%	44% 23%	**
Too few resources allocated to hospital	24% 39%	16% 48%	***

Upper row: % agreement (% of 'definitely yes' or 'yes, probably').
Lower row: % disagreement (% of 'definitely not' or 'probably not').
*: $p<0.05$, **: $p<0.01$, ***: $p<0.001$.

In each survey, a percentage agreement with any individual error cause that fell into the factor on "staff workload" was higher than those with the other items. For instance, "working under great workload" was perceived as a dominant cause of medical errors in both surveys. Patients assigned high agreements – higher than 70% – with the other two statements related to staff workload as causes of medical errors in 2007: "there are fewer nurses" and "fewer doctors" work than actually required in a hospital. The percentage agreements with all three individual causes relevant to the staff workload increased significantly from 2003 to 2007. In particular, there was a large increase of "fewer doctors than required" as a perceived error cause in the four year interval.

Regarding error causes that fell into the factor on "staff ability", i.e., "staff does not feel sufficiently responsible their tasks" and "staff is not sufficiently competent", patient agreements with these items were much lower than the "staff workload" items. In addition, unlike the views about staff workload, percentage agreements with the "staff ability" items deceased during the four year period. This decrease (and currently at a low level) in patient agreement with staff ability as well as the increased (and high) agreement with staff workload may reflect an impression of the Japanese public that hospital staff currently work under extremely high workload conditions and the public's acknowledgement of the staff's abilities, which indeed they now see less than before as a cause of medical accidents. So in general, these results suggest in general a positive shift of the Japanese public's views of healthcare professionals.

Similar to patient responses to the "staff ability" items, there was a significant decrease in patient agreement with each of statements related to the factor on "lack of management efforts" except for the item "too few resources allocated to hospitals" from 2003 to 2007. The reduction in agreements with the "management effort" items was moderate, i.e., between 3% and 8%. The largest drop in patient agreement with specific error causes was the item, "hospital managements do too little for prevent errors". The downward trend in patient agreement with "lack of management efforts" as an error cause suggests a small but positive change in patient views of healthcare organisation and management – similar to the above-mentioned small, positive change in views about healthcare staff.

5 Discussion

There is increasing recognition that reactions after a serious adverse event by healthcare staff are perceived as important by patients and their relatives [19], [20]. Patients want healthcare staff to acknowledge that an adverse event has taken place [21]. But it may not be entirely obvious when and if patients and their relatives want an apology – and if they do, in what sense of apology do they want this? In general terms, an apology recognises that a wrong has been made, admits fault, assumes responsibility and expresses a sense of regret or remorse [22]-[24]; and a complete apology will typically be expected to include a promise of avoiding similar faults in the future and, when appropriate, compensation for harm caused [23], [25]. In the list of reactions offered to respondents (Table 1), we therefore included, first, the weak and minimal reaction in terms of explanation of what has gone wrong, followed by a mere "partial" apology – an, expression of sympathy with the patient. Only the third option in our proposed options is a genuine apology (following Tavuchis' [22] minimum elements) – an acknowledgement that an error was made for which the fictitious doctor or nurse apologises and for which *responsibility* will be taken. To investigate the effects of compensation and explicit promise to undertake preventive measures these options were included in the final options.

Scher and Darley [23] reported that each of the elements mentioned – acknowledgment, expression of sympathy, taking responsibility – independently contribute to the effectiveness of the apology, and that any type of apology, even one that merely expresses sympathy may be better than nothing.

Similarly, positive effects have been reported for offering exemption of expenses incurred by the adverse event [26] – which have therefore suggested our choice of including in our options for respondents an offer of exemption with and without apology. While the "offer of fee exemption" in the questionnaire did not – for the sake of brevity – explicitly include an "explanation of the event", the automatic presumption among Japanese respondents will almost invariably be that offers of fee exemption do not come out of the blue, so to speak, so the offer will have been accompanied by some explanation about what has happened, i.e., a description that makes the offer of fee exemption plausible or natural. Nevertheless, it is somewhat striking that respondents reject "no apology but fee exemption" to a greater extent than "no apology but explanation" (option a and d in Table 1 and Fig. 1).

Still, the chief result stands: patients want a full apology (regret and responsibility taken) and on top of that they tend to prefer a pledge of preventive actions above personal compensation.

6 Conclusion

The present paper has included dual themes in patient perceptions in Japan of healthcare risk management. For one of these themes, we aimed at uncovering patient attitudes to healthcare staff giving various types of apology after an adverse event. The other theme is related to the recent increase in public concern with patient safety as well as the intensified efforts by hospitals to manage patient safety and to be seen as making an effort to do so.

The major results related to the first theme of the present study are that the most effective reaction as seen by patients after an adverse event is a full apology that includes an explicit statement of apology and admittance of hospital responsibility for the event and a promise of future preventive actions. Nevertheless, this apology was not enough for more than a third (35%) of respondents, who still would not want to return to a clinic or hospital where they had suffered a severe outcome event (17% for the mild outcome).

The results also show that it has little effect on mollifying patients after an adverse event if they are offered exemption of additional expenses for treatment and hospitalisation, unless this is accompanied by an unreserved apology. These results may indicate that patients place the greatest importance on a full apology, followed by their receiving assurance that the hospital will seek to learn from the adverse event. At the same time, our results chime in well with those found by Robbenolt [25] who showed that a partial apology may be worse than no apology. Respondents in our study show that an offer of fee exemption is even less attractive than a simple explanation of what went wrong. Patients possibly view an offer of expense exemption with no apology as a demonstration of arrogance by the hospital – as if the adverse event could be nullified simply by money with no admittance of responsibility.

Differences in strength or weakness of patient acceptance of staff apology were also found between respondents when grouped by age, gender, last 24-month experience of hospitalisation, and experience of having suffered medical errors. Patterns in differences in responses among patient groups were similar to those found when asking about patient expectations to a doctor's disclosure after a medical error. For instance, inpatients and patients who had recent experience of hospitalisation exhibited more positive attitudes than patients having no recent hospitalisation in terms of both expectations to a doctor's disclosure and acceptance of a doctor's apology. An opposite trend was found when comparing responses between female and male respondents: on the one hand, female patients exhibited more positive expectations to staff actions after the adverse event than male, whereas a smaller percentage of women accepted staff apology than men regardless of the kind of apology. This discrepancy may be influenced by national culture – for instance, Japan is characterised as strong masculinity in the femininity-masculinity dimension made known by Hofstede [27], including life style, gender roles, and gender difference in personality [28], [29].

In the present questionnaire, we did not use a question that directly asked patients about their acceptance of a given staff apology, for instance, whether they forgive the doctor or hospital or whether they would be satisfied. Instead, in order to acquire patient responses that reveal their level of acceptance or satisfaction reliably but indirectly, we asked them about the likelihood of their using the same clinic or hospital again. This type of question is natural in the context of the current healthcare system in Japan that includes free-access to any hospital or clinic and which has a 70% coverage of expenses by the national health insurance. In contrast, it would be less natural to use this question technique in countries that have different healthcare systems. For instance, in several countries in Europe where patients will often accept a public hospital to which they will typically be referred by their general practitioner or specialist.

With regard to the second theme of this study, we compared results from the present survey sample with those obtained in a previous survey conducted in 2003 in terms of not only patient expectations to a doctor's actions after an adverse event but also their views of other issues related to patient safety. The general trend of changes in the four-year interval can be summarised as follows: Japanese patients have gained more positive views of healthcare professionals and slightly more positive views of healthcare organisations in so far that their expectations of staff actions taken after an adverse event has become more positive and their awareness of staff workload has become greater, their wish to have staff ability checked less pronounced, and their view of management efforts as causal factors of medical errors has become less critical. Based on these findings, we conclude the hypothesis of a "positive change" in patient views mentioned in the Introduction was supported. However, compared with the levels of Danish respondents in 2003 [11], Japanese patients were still more suspicious about healthcare staff attitudes to error reporting and interaction with patient after the adverse event. Therefore, we would suggest that further efforts may possibly improve patient views and attitudes even further in a positive directions vis-à-vis hospitals and hospital professionals in Japan.

Finally, as mentioned above, the present levels in Japan of patient attitudes to healthcare professionals and organisations may not be regarded as entirely satisfactory. However, from the results obtained in this study, it is evident that many patients have "realistic" views and a well-informed recognition of patient safety issues. For instance, most of the patients agreed with the statement that anyone can make a mistake, and many disagreed that a ward or department that reports few errors can also be expected to have few errors. Also, inpatients and patients who had a recent experience of hospitalisation exhibited more positive views of healthcare staff and organisation. This respondent group makes more interactions and exchanges richer communications with healthcare professionals – so in fact, direct experience of how the healthcare system works makes for a more positive view than mere hearsay. Therefore, hospital leaders and managers might consider adopting a "visualisation policy" for hospitals, making risk management and their activities more visible to patients, families and people outside of the organisation. Finally, we suggest that a mature healthcare risk management policy should include guidelines for the healthcare staff – and in serious cases, hospital managers – issuing a sincere apology to any patient who has suffered injury after an avoidable adverse event as soon as possible. The implications suggested in this study were derived from the survey results of Japanese respondents. But the relations between patient and healthcare staff do not seem to differ largely among most western countries, so we believe that the same recommendations would be applicable not only in Japan but also in many of other countries.

Acknowledgments. This work was in part supported by Grant-in-Aid for Scientific Research A(2) (No. 18201029), Japan Society for the Promotion of Science, funded to the first author.

References

1. Japan Nursing Association: White paper of nursing, pp. 143–157. Japan Nursing Association Press, Tokyo (2002) (in Japanese)
2. Deguchi, M.: The Present States of Medical Accidents Analysed from Press Reports in Japan. In: Annual Report 2005, pp. 191–197. Japan Medical Research Institute, Tokyo (2006) (in Japanese)
3. Taylor-Adams, S., Vincent, C., Stanhope, N.: Applying Human Factors Methods to the Investigation and Analysis of Clinical Adverse Events. Safety Science 31, 143–159 (1999)
4. Vincent, C., Stanhope, N., Crowley-Murphy, M.: Reasons for Not Reporting Adverse incidents: An Empirical Study. Journal of Evaluation in Clinical Practice 5(1), 13–21 (1999)
5. Gallagher, T.H., Waterman, A.D., Ebers, A.G., Fraser, V.J., Levinson, W.: Patients' and Physicians' Attitudes regarding the Disclosure of Medical Errors. Journal of the American Medical Association 289(8), 1001–1007 (2003)
6. Hingorani, M., Wong, T., Vafidis, G.: Patients' and Doctors' Attitudes to Amount of Information Given after Unintended Injury during Treatment: Cross Sectional, Questionnaire Survey. British Medical Journal 318, 640–641 (1999)
7. Hobgood, C., Peck, C.R., Gilbert, B., Chappell, K., Zou, B.: Medical Errors – What and When: What Do Patients Want to Know? Academic Emergency Medicine 9(11), 1156–1161 (2002)
8. Witman, A.B., Park, D.M., Hardin, S.B.: How Do Patients Want Physicians to Handle Mistakes? Archives of Internal Medicine 156, 2565–2569 (1996)
9. Tayeb, M.: Conducting Research across Cultures: Overcoming Drawbacks and Obstacles. International Journal of Cross Cultural Management 1(1), 91–108 (2001)
10. Helmreich, R.L.: Culture and Error in Space: Implications from Analog Environments. Aviation, Space, and Environmental Medicine 71(9-11), 133–139 (2000)
11. Itoh, K., Andersen, H.B., Madsen, M.D., Østergaard, D., Ikeno, M.: Patient Views of Adverse Events: Comparisons of Self-reported Healthcare Staff Attitudes with Disclosure of Accident Information. Applied Ergonomics 37, 513–523 (2006)
12. Andersen, H.B., Madsen, M.D., Ruhnau, B., Freil, M., Østergaard, D., Hermann, N.: Do Doctors and Nurses Know What Patients Want after Adverse Events? In: 9th European Forum on Quality Improvement in Health Care, Copenhagen, Denmark (May 2004)

13. Itoh, K., Abe, T., Andersen, H.B.: A Questionnaire-based Survey on Healthcare Safety Culture from Six Thousand Japanese Hospital Staff: Organisational, Professional and Department/Ward Differences. In: Tartglia, R., Bagnara, S., Bellandi, T., Albolino, S. (eds.) Healthcare Systems Ergonomics and Patient Safety: Human Factor, a Bridge between Care and Cure, pp. 201–207. Taylor & Francis, London (2005); (Proceedings of the International Conference on Healthcare Systems Ergonomics and Patient Safety, HEPS 2005, Florence, Italy, March-April 2005)
14. Robbenolt, J.K.: Apologies and Medical Error. Clinical Orthopaedics and Related Research 467(2), 376–382 (2009)
15. Kraman, S.S., Hamm, G.: Risk management: Extreme honesty be the best policy. Annals of Internal Medicine 131, 963–967 (1999)
16. Andersen, H.B., Madsen, M.D., Hermann, N., Schiøler, T., Østergaard, D.: Reporting Adverse Events in Hospitals: A Survey of the Views of Doctors and Nurses on Reporting Practices and Models of Reporting. In: Johnson, C. (ed.) Proceedings of the Workshop on the Investigation and Reporting of Incidents and Accidents, Glasgow, UK, July 2002, pp. 127–136 (2002)
17. Campbell, J.L., Ramsay, J., Green, J.: Age, Gender, Socioeconomic, and Ethnic Differences in Patients' Assessments of Primary Health Care. Quality in Health Care 10, 90–95 (2001)
18. Itoh, K., Andersen, H.B.: Causes of Medical Errors as Perceived by Patients and Healthcare Staff. In: Aven, T., Vinnem, J.E. (eds.) Risk, Reliability and Societal Safety. Specialisation Topics, vol. 1, pp. 179–185. Taylor & Francis, London (2007); (Proceedings of the European Safety and Reliability Conference 2007 — ESREL 2007, Stavanger, Norway (June 2007))
19. Leape, L.L.: Understanding the power of apology: How saying "I'm sorry" helps heal patients and caregivers. Focus on Patient Safety: A Newsletter from the National Patient Foundation 8(4), 1–3 (2005)
20. Madsen, M.D.: Improving Patient Safety: Safety Culture and Patient Safety Ethics. DTU Risoe, Roskilde, Denmark (2006), `http://www.risoe.dk/rispubl/SYS/syspdf/ris-phd-25.pdf`
21. Manser, T., Staender, S.: Aftermath of an adverse event: supporting health care professionals to meet patient expectations through open disclosure. Acta Anaesthesiologica Scandinavica 49, 728–734 (2005)
22. Tavuchis, N.: Meta Culpa: A Sociology of Apology and Reconciliation, pp. 15–41. Stanford University Press, Stanford (1991)
23. Scher, S.J., Darley, J.M.: How Effective Are the Things People Say to Apologize? Effects of the Realization of the Apology Speech Act. Journal of Psycholinguistic Research 26, 127–140 (1997)
24. Gill, K.: The moral functions of an apology. Philosophical Forum 31, 11–27 (2000)
25. Robbenolt, J.K.: Apologies and Legal Settlement: An Empirical Examination. Michigan Law Review 102(3), 460–516 (2003)
26. Cohen, J.R.: Advising Clients to Apologize. Southern California Law Review 72, 1009–1069 (1998)
27. Hofstede, G.: Cultures and Organizations: Software of the Mind. McGraw-Hill, New York (1991)
28. Feingold, A.: Gender Differences in Personality: A Meta-Analysis. Psychological Bulletin 116, 429–456 (1994)
29. Hyde, J.S.: The Gender Similarity Hypothesis. America Psychologist 60(6), 581–592 (2005)

A Cross-National Study on Healthcare Safety Climate and Staff Attitudes to Disclosing Adverse Events between China and Japan

Xiuzhu Gu and Kenji Itoh

Dept. of IE and Management, Tokyo Institute of Technology
2-12-1 Oh-okayama Meguro-ku, Tokyo 152-8552, Japan
{xiuzhu.g.aa,itoh.k.aa}@m.titech.ac.jp

Abstract. The present paper reports comparative results of safety climate in healthcare and staff attitudes to error reporting and interaction with patients between China and Japan. Using two language versions of questionnaire, we collected response data from hospital staff in China (in 2008) and Japan (in 2006). Significant differences were observed in most dimensions of safety climate between these two countries, though not in the same direction in terms of positive or negative nature. In contrast, there was a uniform national difference in staff attitudes to error reporting. Chinese doctors and nurses being significantly less willing than their Japanese colleagues to engage in any action or interaction with patients after an adverse event, regardless of the severity of the event. Finally, we discuss possible sources of these differences in safety climate and staff attitudes between the two countries, and some implications for improving healthcare safety climate.

Keywords: Healthcare safety climate; incident reporting; adverse event; cross-national survey.

1 Introduction

As in western countries and Japan, there has been a rapid increase of concerns with patient safety in China. Safety culture has been widely recognized as being of major importance to patient safety [1], and therefore a number of studies have been carried, aiming at developing safety culture tools or instruments and applying questionnaire surveys to measuring and diagnosing a specific healthcare organization as well as at the nation level in these countries – e.g., in the USA [2] and Europe [3]. In recent years, safety culture has also drawn great attention in China. A sign of the increasing awareness of the importance of safety culture is that the Chinese government has established strategic goals for improving patient safety in 2008, including a non-punitive management style to encourage voluntary incident reporting in healthcare organizations as one of ten major goals [4]. While in recent years, a small number of studies have appeared of safety culture in domains outside healthcare in China– e.g. [5] [6], a few other studies have also been made in healthcare – perhaps inspired by the strategic goals laid down by the Chinese government as mentioned [7] [8].

The original version of this chapter was revised: The copyright line was incorrect. This has been corrected. The Erratum to this chapter is available at DOI: 10.1007/978-3-642-11750-3_10

P. Palanque, J. Vanderdonckt, and M. Winckler (Eds.): HESSD 2009, LNCS 5962, pp. 44–53, 2010.

But no comprehensive, systematic survey has been carried out yet in the healthcare sector. For instance, among the few Chinese studies relevant to this field, a safety culture assessment was recently published reporting on nurse responses to a small questionnaire having 19 items [8].

Given the background mentioned above, it is an urgent goal to uncover safety culture in Chinese healthcare. In particular, to highlight its important characteristics, cross-national studies may be required, comparing with other country samples using the same questionnaire. Motivated by these requirements, we conducted a safety culture survey in China, collecting data by applying a Chinese translation of a questionnaire which was used for a Japanese survey [9]. Comparative analyses were performed using the two country samples, aiming at identifying the impacts of different healthcare systems and safety regulations as well as national culture on safety culture or climate. The terms "safety culture" and "safety climate" suggest somewhat different approaches to studying and modeling the shared understanding that members of an organization or group may have: "culture" is usually meant to refer to the long-term and relatively unchanging and mostly tacit values, norms and assumptions of a group, whereas "climate" is usually meant to refer to the more context-based and more changeable ones [10] [11]. Hereafter we use the term, safety climate with its emphasis on the local, changeable nature and explicit attitudes and perceptions.

In the present paper, we report results of cross-national analysis between China and Japan in healthcare safety climate and staff attitudes to error reporting and interactions with the patient after an adverse event. We discuss possible reasons or factors contributing to variations of the safety climate between these two countries. Based on the discussions, we also explore some implications for safer healthcare systems in the near future.

2 Questionnaire

The questionnaire used in this study was a Chinese translation of one used in a Japanese survey [12]. The Chinese version of questionnaire was thoroughly checked from professional points of view by several doctors and nurses working in the hospital from which data were collected as well as consistency of the two language versions. The questionnaire comprised four sections, among which this paper focused on the first two sections, as well as a demographic section.

The first section was comprised of 57 question items to elicit respondents' perceptions of their job, hospital management, and factors that might impact on safety performance. Respondents were asked to rate their agreement or disagreement with each statement on a five-point Likert type scale, ranging from 'strongly disagree' to 'strongly agree'. This section was originally adapted from the instrument used for "Operating Team Resource Management Survey" developed by Robert Helmreich and his research group [13].

In the second section, respondents were asked about their behavior and actions in terms of reporting own errors and interaction with patients that have been victims of such errors. Respondents' reactions were elicited as responses to three fictitious incident cases (vignettes) having different outcome severity: Cases A (near-miss), B (mild outcome) and C (severe outcome). The respondents were asked to read each case and subsequently rate the likelihood of engaging in various actions described in the questionnaire. The likelihood rating was also made on a five point Likert-type scale ranging from 'definitely no' to 'definitely yes'. The incident cases and questions were reproduced from those of the Danish Patient Safety Questionnaire [14].

The Chinese survey was carried out between October and December 2008, collecting data from doctors, nurses, pharmacists and other professionals working in a university hospital – having approximately 850 beds, 650 doctors and 700 nurses working – located in Shanghai, China. Responses and response rates of each professional group are shown in Table 1, which also includes the response information of the Japanese survey [9] which was used for cross-national analysis in this study.

Table 1. Number of responses used for cross-national analysis

Professional groups	China		Japan [9]	
	Number of responses	Response rate	Number of responses	Response rate
Doctor	388	78%	1,005	51%
Nurse	611	87%	17,858	88%
Pharmacist	37	74%	542	48%
Others/NA	20	40%	2,261	-
Total	1,056	81%	21,666	84%

3 Results of Cross-National Analysis

3.1 Healthcare Safety Climate

As a framework for cross-national analysis of safety climate, we employed a twelve dimensional model [9], which was derived by applying the principal component analysis to the Japanese sample of the 2006 survey. Table 2 describes the meaning of each safety climate dimension, and Cronbach's α is also shown when applying the model to the Chinese sample. As can be seen in this table, Cronbach's α was very low for some dimensions. Therefore, we compared the two countries' safety climate in terms of dimensions having higher than 0.4 of Cronbach's α. The threshold of 0.4 is unusually low, and the standard threshold of α for accepting a construct or factor is 0.7 [15]. However, it may be argued that satisfactory levels of alpha depend on test use and interpretation. Even relatively low levels of criterion reliability do not seriously attenuate validity coefficients [16]. A lower threshold may be applied, especially with factors that comprise a small number of items [17]. Still, we do not propose to apply a lower than usual threshold for identifying the reliability of a construct. For the rest of dimensions, we made item-based comparisons between the Chinese and the Japanese sample for a few typical or representative items to a specific safety climate dimension, e.g., "human error is inevitable" and "errors are a sign of incompetence" for the dimension on recognition of human error.

Table 2. Dimensions of safety climate for comparisons between China and Japan [9]

Safety Climate dimensions (Cronbach's α)	Description
I. Recognition of communication (α = 0.700)	Acknowledgement of importance of communication and coordination during performing a job within an organization or among team members
II. Morale and motivation (α = 0.699)	Employees' morale and motivation for work and organization
III. Power distance (α = 0.675)	Psychological distance between leaders or superiors and subordinate members. In a large power distance organization there may be bureaucratic, authoritative atmosphere, limited open communication between leaders and their subordinates within a workplace and lack of communication between departments
IV. Recognition of stress effects on own performance. (α = 0.327)	Realistic understanding and recognition about effects of stress, fatigue and other psychological factors on their own work performance.
V. Trust in management (α = 0.288)	Staff trust in hospital management systems, senior managers, and leaders/superiors in their department or organization
VI. Safety awareness (α = 0.389)	Staff members' awareness of safety for their jobs and patient safety issues. High awareness members are likely to take risk avoidance behavior/attitudes during task performance
VII. Awareness of own competence (α = 0.630)	Staff members' awareness of their own competence and skills. High awareness members believe staff competence and skills are the most important factors for working in a hospital
VIII. Collectivism- individualism (α = 0.282)	Proportion of staff members taking team-oriented or collectivistic behavior in an organization
IX. Cooperativeness (α = 0.055)	Proportion of staff members taking cooperative attitudes and behaviors within an organization
X. Recognition of stress management for team members (α = 0.534)	Realistic acknowledgement or awareness about other team members' stress and fatigue levels while they are working by team
XI. Seniority dependency (α = 0.435)	Tend to depend on seniors or seniority system for work management in an organization.
XII. Recognition of human error (α = 0.050)	Realistic acknowledgement of human errors within a department/organization

According to this procedure of the cross-national comparisons, Table 3 shows percentage agreements and disagreements of both doctor and nurse groups in China and Japan for each of the selected dimensions. The percentage [dis]agreement is referred to as a proportion of a specific respondent group, e.g., Chinese doctors and Japanese nurses, that strongly or slightly [dis]agreed with the items comprised in a given dimension. This index is computed as follows: before the mean score calculation, figures of responses were reversed for an item which has a negative meaning to a dimension label, i.e., response rank "1" was changed to "5", and vice versa. A possible range of the mean score (i.e., 1.0–5.0) was divided into five ranks, the top two ranks, 3.41–5.00, are allocated to an 'agreement' class, and the last two, 1.00–2.60, to a 'disagreement'. Table 3 also includes significance levels computed by the Mann-Whitney test between Chinese and Japanese staff for each dimension of safety climate.

As can be seen from Table 3, there were significant differences between Chinese and Japanese doctors for all dimensions of safety climate except for morale and motivation. For some dimensions, the safety climate elicited from Chinese doctors was more negative than those of Japanese: Chinese doctors' recognition of communication and stress management was weaker than that of the Japanese; and the power distance was much larger in the Chinese hospital. In contrast, Chinese doctors were more aware of their own skills and competence, and their dependency on senior staff and a seniority system was significantly weaker than Japanese.

As results of item-based comparisons of staff responses, significant differences were observed for most of statements, i.e., 49 out of 57 items for doctors and 51 items for nurses, between the two countries. Regarding error-related issues, Chinese doctors showed themselves less realistic recognition of human errors than Japanese. A greater number of doctors agreed with the statement "errors are a sign of incompetence" in China than Japan ($p<0.001$, agreement rate of Chinese and Japanese were 23% and 7%, respectively), and disagreed with the item "human error is inevitable" ($p<0.001$; disagreement: 51% and 10% – almost all Japanese doctors accepted this statement).

Table 3. Safety climate comparisons between Chinese and Japanese healthcare staff

Safety climate dimensions	China		Japan		*p* (Mann-Whitney)	
	Dr.	Ns.	Dr.	Ns.	Dr.	Ns.
I. Recognition of communication	73% 4%	76% 2%	95% 0%	95% 0%	0.000	0.000
II. Morale & motivation	60% 8%	65% 7%	65% 9%	47% 19%	0.192	0.001
III. Power distance	17% 39%	19% 41%	1% 84%	1% 82%	0.000	0.000
VII. Awareness of own competence	73% 2%	75% 2%	50% 6%	32% 12%	0.000	0.000
X. Recognition of stress mgt. for members	72% 3%	70% 4%	84% 1%	81% 1%	0.005	0.000
XI. Seniority dependency	67% 6%	57% 6%	85% 1%	55% 7%	0.000	0.051

Upper row: % agreement, Low row: % disagreement.

Similar to doctors' responses – but with a slightly different trend – Chinese nurses exhibited significantly negative views of or attitudes to some dimensions of safety climate compared to their Japanese colleagues, i.e., weaker recognition of communication, larger power distance, and weaker recognition of stress management for team members. As can be seen from Table 3, however, Chinese nurses revealed more positive reactions to two dimensions than Japanese: higher morale and motivation, and much stronger awareness of their own competence – the difference for the latter dimension was identical to that of the doctor group. Similar differences to those between Japanese and Chinese doctors were also found in the nurse samples. But because of the low reliability of the constructs (low alpha) the comparisons were made on an item comparison basis. Again, we found a much less realistic recognition of errors by Chinese nurses compared to their Japanese colleagues.

3.2 Staff Attitudes to Error Reporting

Using responses to the second section of the questionnaire, we analyzed cross-national differences between China and Japan in healthcare staff attitudes to error reporting (for Cases A to C) and interactions with patients (for Cases B and C) based on professional groups. Large differences were observed between the two countries for both doctor and nurse attitudes, and characteristics of cross-national differences were shared between the doctor and the nurse group. The comparative results of the two country samples are shown for the doctor and nurse group respectively in Table 4 and 5, based on three cases of outcome severity in terms of percentage agreement and disagreement (strongly or slightly) with each action.

Table 4. Attitudes to error reporting comparisons between Chinese and Japanese doctors

Actions	Case	China		Japan		*p*
		% agr.	% disag.	% agr.	% disag.	
Keep the event secret	A	30%	43%	7%	83%	0.000
	B	29%	40%	3%	87%	0.000
	C	24%	42%	2%	94%	0.000
Report to leader or doctor in charge	A	48%	22%	78%	9%	0.000
	B	61%	13%	91%	3%	0.000
	C	61%	12%	96%	1%	0.000
Write in patient's case record	B	28%	37%	76%	9%	0.000
	C	30%	29%	87%	4%	0.000
Report to the local system	A	30%	29%	66%	16%	0.000
	B	38%	23%	73%	7%	0.000
	C	42%	20%	93%	2%	0.000
Inform patient of the error and risk	B	25%	39%	52%	17%	0.000
	C	36%	32%	92%	2%	0.000
Admit own error	B	27%	38%	52%	17%	0.000
	C	36%	31%	86%	4%	0.000
Apologize to patient	B	40%	31%	62%	14%	0.000
	C	43%	24%	89%	3%	0.000

% agr. : % agreement (% of respondents definitely or slightly agree with the statement).
% disag. : % disagreement (% of respondents definitely or slightly disagree with the statement).

Table 4 showed that the attitudes of Chinese and Japanese doctors are, in all the three cases, highly significantly different. Chinese doctors had weaker willingness to report the error, and they also showed themselves less likely to inform about the error or apologize to the patient than Japanese doctors. With the case's severity increased, Japanese doctors showed higher willingness to report the error and do interaction with the patient, for example, inform patient of the error and risk; while Chinese doctors'

Table 5. Attitudes to error reporting comparisons between Chinese and Japanese nurses

Actions	Case	China % agr.	China % disag.	Japan % agr.	Japan % disag.	p
Keep the event secret	A	23%	47%	8%	79%	0.000
	B	28%	43%	1%	94%	0.000
	C	26%	41%	1%	92%	0.000
Report to leader or doctor in charge	A	54%	18%	80%	9%	0.000
	B	65%	9%	93%	3%	0.000
	C	63%	9%	92%	4%	0.000
Write in patient's case record	B	28%	32%	73%	7%	0.000
	C	29%	28%	71%	6%	0.000
Report to the local system	A	36%	21%	71%	13%	0.000
	B	38%	18%	91%	3%	0.000
	C	42%	15%	90%	3%	0.000
Inform patient of the error and risk	B	27%	32%	43%	14%	0.000
	C	35%	31%	71%	5%	0.000
Admit own error	B	33%	30%	44%	14%	0.000
	C	37%	30%	57%	9%	0.000
Apologize to patient	B	44%	26%	66%	6%	0.000
	C	46%	25%	76%	4%	0.000

% agr. : % agreement (% of respondents definitely or slightly agree with the statement).
% disag. : % disagreement (% of respondents definitely or slightly disagree with the statement).

willingness doesn't increase so much. About error reporting, both Chinese and Japanese doctors prefer reporting to leader or doctor in charge than other ways.

Similar to doctors, there were also highly significant differences between Chinese and Japanese nurses in their attitudes for all error reporting actions and reactions with the patient regardless of the levels of outcome severity. As shown in Table 5, Chinese nurses exhibited more negative attitudes to following a policy of openness than their Japanese colleagues for every action offered in the questionnaire, e.g., they were more likely to keep the event to themselves; less willing to report the error to their leader or doctor in charge; less willing to submit an incident report to the local system; less likely to inform the patient about the event and future health risk; more reluctant to admit their own errors; and more reluctant to apologize to the patient.

Conversely, almost all Japanese nurses indicated that they would report the event to their leader or a doctor in charge, and submit it to their local system for the two cases involving injury, i.e., Cases B and C, and they were significantly more likely to do so than for the non-injured near-miss case A, while there were no large differences in their willingness to take these reporting actions between the mild and the severe outcome case. However, their interactions with the patient, e.g., apology to the patient, informing the patient about the event and the future risk, and admittance of their own errors, were more positively affirmed in the severe than in the mild outcome case. In contrast, among the Chinese nurses no large differences were observed between the cases involving injury, i.e., B and C, in their willingness to do error reporting or to interact with the patient.

4 Discussion: Sources of Cross-National Differences

In the present paper, we applied the safety climate construct derived in a previous study [9] to a Chinese sample as an analysis framework for making cross-national comparisons. This application of the Japanese model seems to be a major reason why Cronbach's α calculated were very low for some dimensions, in particular, for recognition of human error, cooperativeness and recognition of stress effects on one's own performance. All of these dimensions are closely connected to issues concerning human factors at work and their effects on work performance. Here we refer to 'human factors' as the collection of knowledge about how humans perform, cognitively, socially and physiologically, in work contexts and especially contexts involving technological systems and safety critical operations. The low internal reliability of these dimensions may indicate that Chinese respondents understand the meaning of the component items in a slightly different way than intended and different from the way in which the Japanese respondents understand them. This difference in understanding the items may have to with the lack of familiarity that the Chinese respondents have with notions about performance shaping factors such as the effects of fatigue or poor communication on performance. This speculation is supported by the fact that the Chinese healthcare staff is unfamiliar with safety training in the style of CRM (crew resource management) [13], and are thus not familiar with human factors as described above and especially with human cognitive limitations in work performance.

Based on the cross-national analyses with the Japanese data, it appears as important characteristics of safety climate in Chinese healthcare that doctors' recognition of communication for safety as well as the role of stress management for team members is very weak. It may be the reason why improving the effective communication among healthcare staff was required to include as one of important patient safety goals established by the government in 2008 [4]. At the same time, power distance was larger and the doctors' recognition of human error issues was not enough realistic. For instance, a number of Chinese doctors disagreed strongly with the statement, "human error is inevitable". This point of view is presumably based on their concepts of human error mechanism and possibly derived by an idea of the "true professional" as being infallible; and again, this is presumably linked to the fact that they have never received any training in human performance for safety critical tasks.

A larger power distance may be conjectured to be a major source of the greater reluctance of Chinese staff to reporting errors and adverse events. According to interviews with risk managers in the hospital surveyed, healthcare staff would receive punishment when they were found to make an error. This finding may be supported by the result identified by Liu et al. [8], that there was a punitive safety culture in hospitals in China. In addition, patients in China have a possibility to initiate complaint procedures and apply for compensation – a course of action which is not open to Japanese patients in the same way. Again, the threat of punishment and complaints would naturally lead Chinese doctors and nurses to be very reluctant to report their own errors or be in any way open about them.

When confronted with a case involving an error leading to a serious outcome, Chinese staff became slightly more willing to report the error and to interact with the patient. In contrast, Japanese nurses expressed a higher willingness to report the case if it involved a milder outcome rather than a severe one; they also showed themselves more reluctant to interact with patient – offer an apology to the patient or admit their own error – than to report the event to their leader or to the local system. However, it may well be the case that the chief reason why the Chinese respondents express a greater willingness to report the severe incident than the minor one is because the error is obvious and impossible to hide. A major reason for consistent levels of error reporting attitudes expressed by Japanese nurses between the two cases involved injury seems to be based on a widespread safety practice or custom within the nursing department: nurses are always pushed by their leaders and senior staff to report their errors, and in fact are strongly encouraged to do so ('must' do so in many hospitals) when they involve in an injured event – even very minor one – although incident reporting is formally operated voluntarily in most Japanese hospitals.

5 Conclusion

This study identified some important characteristics of healthcare safety climate in China by comparing with the Japanese sample: Chinese staff put much less weight on the importance of communication among staff members than Japanese, and they showed less realistic recognition of human performance and human error issues. Also, larger power distance was identified in the Chinese healthcare settings. In contrast, Chinese nurses' morale and motivation as well as their awareness of own competencies were significantly higher than those of Japanese nurses. Chinese staff was more likely to have negative attitudes to error reporting and to interaction with patients who suffered an adverse event than Japanese staff.

From the results and discussions obtained in this paper, we would suggest that safety training particularly involving human factors aspects should be required to foster a positive safety climate in China. Negative staff attitudes to adverse event reporting may be primarily created by fear of sanctions and disrepute. Behind the reluctance to report errors - and thus be able to learn from these – there is, we suggest a blame culture affecting Chinese healthcare, as indeed also suggested by the hospital risk managers through our interviews. A large power distance observed in the Chinese sample might also play a role in discouraging openness and adverse event reporting. Therefore, we would suggest, as one of the urgent hospital-wide initiatives in China, to steer risk management in the direction of a non-punitive style, and to adopt safety training, particularly including learning from errors and accidents. We believe this in turn would contribute to establishing an effective learning culture in healthcare and thus to greater patient safety.

Acknowledgments

This work was in part supported by Grant-in-Aid for Scientific Research A (2) (No. 18201029), Japan Society for the Promotion of Science. We would like to acknowledge Henning Boje Andersen, Senior Scientist in Technical University of Denmark for his insightful discussions and comments. We are also grateful to the risk manager and risk management personnel in the hospital surveyed for supporting data collection and valuable information about their risk management procedures and systems.

References

1. Itoh, K., Andersen, H.B., Madsen, M.D.: Safety Culture in Healthcare. In: Carayon, P. (ed.) Handbook of Human Factors and Ergonomics in Healthcare and Patient Safety, pp. 199–216. Lawrence Erlbaum Associates, Mahwah (2006)
2. Kohn, L.T., Corrigan, J.M., Donaldson, M.S. (eds.): To Err Is Human: Building a Safer Health System. National Academy Press, Washington (1999)
3. Department of Health: An Organisation with a Memory: Report of an Expert Group on Learning from Adverse Events in the National Health Service Chaired by the Chief Medical Officer, Stationery Office, London (2000)
4. Chinese Hospital Association, China (2009), `http://www.cha.org.cn/GD/GeneralDocument/GDContent.aspx?ContentId=608&ClassId=167&ChannelId=38` (accessed 14 September)
5. Lin, S., Tang, W., Miao, J., Wang, Z., Wang, P.: Safety Climate Measurement at Workplace in China: A Validity and Reliability Assessment. Safety Science 46, 1037–1046 (2008)
6. Ma, Q., Yuan, J.: Exploratory Study on Safety Climate in Chinese Manufacturing Enterprises. Safety Science 47(7), 1043–1046 (2009)
7. Wu, X., Piao, Y., Fang, X.: Investigation on Nurses' Perception on Hospital Safety Culture. Journal of Nursing Science 24, 7–9 (2009) (in Chinese)
8. Liu, Y., Kalisch, B.J., Zhang, L., Xu, J.: Perception of Safety Culture by Nurses in Hospitals in China. Journal of Nursing Care Quality 24(1), 63–68 (2008)
9. Itoh, K., Andersen, H.B.: A National Survey on Healthcare Safety Culture in Japan: Analysis of 20,000 Staff Responses from 84 Hospitals. In: Proceedings of the International Conference on Healthcare Systems Ergonomics and Patient Safety, HEPS 2008, Strasbourg, France (June 2008) (CD-ROM)
10. Guldenmund, F.W.: The Nature of Safety Culture: A Review of Theory and Research. Safety Science 34(1-3), 215–257 (2000)
11. Madsen, M.D., Andersen, H.B., Itoh, K.: Assessing Safety Culture and Climate in Healthcare. In: Carayon, P. (ed.) Handbook of Human Factors and Ergonomics in Healthcare and Patient Safety, pp. 693–713. Lawrence Erlbaum Associates, Mahwah (2007)
12. Itoh, K., Abe, T., Andersen, H.B.: A Questionnaire-based Survey on Healthcare Safety Culture from Six Thousand Japanese Hospital Staff: Organisational, Professional and Department/Ward Differences. In: Proceedings of the International Conference on Healthcare Systems Ergonomics and Patient Safety, HEPS 2005, Florence, Italy, March-April 2005, pp. 201–207 (2005)
13. Helmreich, R.L., Merritt, A.C.: Culture at Work in Aviation and Medicine: National, Organizational and Professional Influences. Ashgate, Aldershot (1998)
14. Andersen, H.B., Madsen, M.D., Hermann, N., Schiøler, T., Østergaard, D.: Reporting Adverse Events in Hospitals: A Survey of the Views of Doctors and Nurses on Reporting Practices and Models of Reporting. In: Johnson, C. (ed.) Proceedings of the Workshop on the Investigation and Reporting of Incidents and Accidents, Glasgow, UK, July 2002, pp. 127–136 (2002)
15. Bland, J.M., Altman, D.G.: Statistics Notes: Cronbach's Alpha. British Medical Journal 314(7080), 572 (1997)
16. Schmitt, N.: Uses and Abuses of Coefficient Alpha. Psychological Assessment 8(4), 350–353 (1996)
17. Spiliotopoulou, G.: Reliability Reconsidered: Cronbach' Alpha and Paediatric Assessment in Occupational Therapy. Australian Occupational Therapy Journal 56(3), 150–155 (2009)

Cognitive Modelling of Pilot Errors and Error Recovery in Flight Management Tasks

Andreas Lüdtke[1], Jan-Patrick Osterloh[1], Tina Mioch[2], Frank Rister[3], and Rosemarijn Looije[2]

[1] OFFIS e.V., Escherweg 2, 26121 Oldenburg, Germany
[2] TNO Human Factors, Kampweg 5, 3796 DE Soesterberg, The Netherlands
[3] Hapag-Lloyd Flug, Flughafenstrasse 10, 30855 Langenhagen, Germany
{luedtke,osterloh}@offis.de,
{tina.mioch,rosemarijn.looije}@tno.nl, frank.rister@hamburg.de

Abstract. This paper presents a cognitive modelling approach to predict pilot errors and error recovery during the interaction with aircraft cockpit systems. The model allows execution of flight procedures in a virtual simulation environment and production of simulation traces. We present traces for the interaction with a future Flight Management System that show in detail the dependencies of two cognitive error production mechanisms that are integrated in the model: Learned Carelessness and Cognitive Lockup. The traces provide a basis for later comparison with human data in order to validate the model. The ultimate goal of the work is to apply the model within a method for the analysis of human errors to support human centred design of cockpit systems. As an example we analyze the perception of automatic flight mode changes.

Keywords: Human Error Prediction, Human-Centred Design, Cognitive Model.

1 Introduction

Aircraft pilots are faced with a complex traffic environment. Cockpit automation and support systems help to reduce complexity. Currently, a lot of research is done to improve the onboard management of flight trajectories and the negotiation of trajectory changes with Air Traffic Control (ATC). During the flight, many factors may induce changes to the original flight plan, e.g. bad weather, traffic conflicts, or runway changes. In future air traffic management, an aircraft will be equipped with an advanced flight management system that provides information on the current traffic and weather status in an intuitive form. This allows pilots to easily adapt a flight route via a graphical Advanced Human Machine Interface (AHMI). Voice communication between aircraft and ATC will be partly replaced by Data Link communication which provides pilots and controllers with a detailed electronic picture of the time and space (4D) trajectory. This allows efficient negotiation of route changes and improves predictability of conflicts between aircraft or between planned routes and severe weather conditions.

In order to leverage this new air traffic management concept, intuitive and easy-to-use human machine interfaces as well as efficient and robust flight procedures are

---

The original version of this chapter was revised: The copyright line was incorrect. This has been corrected. The Erratum to this chapter is available at DOI: 10.1007/978-3-642-11750-3_10

P. Palanque, J. Vanderdonckt, and M. Winckler (Eds.): HESSD 2009, LNCS 5962, pp. 54–67, 2010.

needed. Safe operation of aircraft is based on normative flight procedures (standard operating procedures) and rules of good airmanship, which we will referred to as 'normative activities'. We define pilot errors as deviations from normative activities. In the past, several cognitive explanations and theories have been proposed to understand why pilots deviate from normative activities (e.g. [7]). The European project HUMAN, in which the research described in this paper is done, strives to pave a way of making this knowledge readily available to designers of new cockpit systems. We intend to achieve this by means of a valid executable flight crew model which incorporates cognitive error-producing mechanisms leading to deviations from normative activities. The model interacts with models of cockpit systems (like advanced flight management systems) in a virtual simulation environment to predict deviations and its potential consequences on the safety of flight. The ultimate objective of HUMAN is to apply this model to analyze human errors and support error prediction in ways that are usable and practical for human-centred design of systems operating in complex cockpit environments.

This paper focuses on the interaction between two highly relevant cognitive error-producing mechanisms:

- routine learning leading to *Learned Carelessness* (effort-optimizing shortcuts leading to inadequate simplifications that omit safety critical aspects of normative activities) and
- attention allocation (deciding where to allocate the limited cognitive resources) leading to *Cognitive Lockup* (failing to switch attention when currently working on a demanding task).

At the initial stage of HUMAN we performed questionnaire interviews with pilots and human factor experts based on a literature survey of error-producing mechanisms. We identified Learned Carelessness and Cognitive Lockup to be among the most relevant mechanisms for modern and future cockpit human machine interfaces.

This paper describes how we modelled these two processes in one integrated executable cognitive flight crew model and discusses in detail hypotheses derived from the model.

2 Re-planning via 4D-Flight Management Systems

Today, the flight management system, which controls the lateral and vertical movement of an aircraft, is operated by a multi-purpose control display unit (MCDU). The MCDU consists of a small monitor and an alphanumerical keyboard, by which the pilots type in the desired flight plan changes. Flight plans consist of a certain number of waypoints, identified by a three or five letter code, which is entered into the MCDU. The airplane's autoflight system can be coupled to the flight plan, which then follows the plan automatically. However, clearance requests and reception for the different sections of the flight plan are mandatory, and are today performed via voice communication with the ATC. Problems with this are that communicating route changes via voice is a lengthy and error-prone process [2], and that the interaction with the MCDU is cumbersome and inefficient (e.g. [6]). As described in the introduction, future flight management systems and their user interfaces try to tackle these problems. For our study we use an advanced flight management system and its AHMI, which have been developed by the German Aerospace Institute (DLR, Braunschweig, Germany). Both systems are used as demonstration settings for the current research, without their design playing a role in the validity of the current research.

The AHMI represents flight plans on a map with their status being graphically augmented by different colours and shapes, e.g. if a new trajectory is generated after the flight plan has been changed, it is displayed as a dotted line, while the active trajectory is solid of another colour (cf. Fig. 1). Still, pilots can insert, move or delete waypoints, but also handle a lot of different events, e.g. display weather radar information, allowing graphical re-planning to avoid a thunderstorm. However, the insertions do not necessarily make use of keyboards such as the MCDU - manipulation is done directly on the map by trackball cursor-control. Any trajectory created by the pilot is generated as a data-link, ready to be sent to ATC for negotiation.

The advanced flight management system and its AHMI is used in HUMAN as a target system to demonstrate the predictive capabilities of the cognitive flight crew model by simulating the interaction between system and crew in different re-planning scenarios according to a set of normative activities.

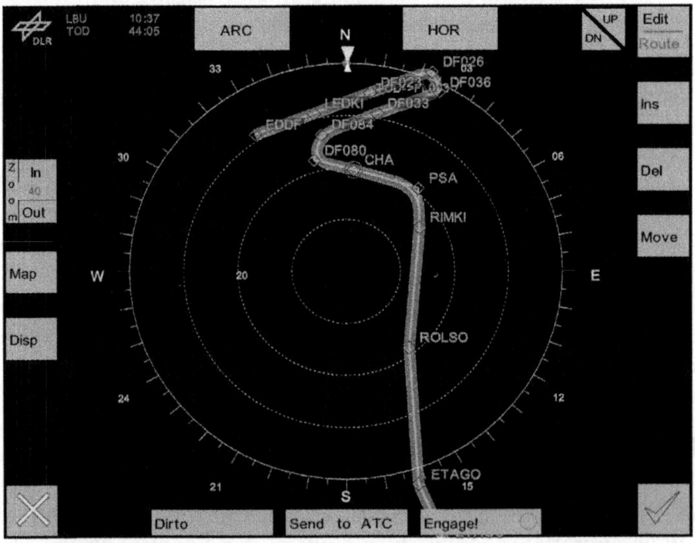

Fig. 1. AHMI of the Flight Management target system

Since this is a new system we had to define the normative activities (NA) from scratch. Knowledge acquisition techniques were used to gather first ideas for the scenarios and NA definition. As a second step, common Standard Operating Procedures (SOP) and Rules of Good Airmanship were the basis of workflow patterns which were applied and refined by test and trainer pilots working in the field of procedure and training-scenario/simulation design.

Next, these procedural workflow patterns were translated into a textual description. This textual description served as the basis for the first plot of NA's in table format. These tables, in turn, were used to model the NAs in the semi-formal task modelling software AMBOSS [20]. The task trees were useful in two ways: first of all, the AMBOSS models were used to reveal the flaws in the NA tables that were undetectable without a simulation. And second, these tables paved the path for a formal model of the normative activities, which are then input for the cognitive architecture. Fig. 2 represents the modelling process.

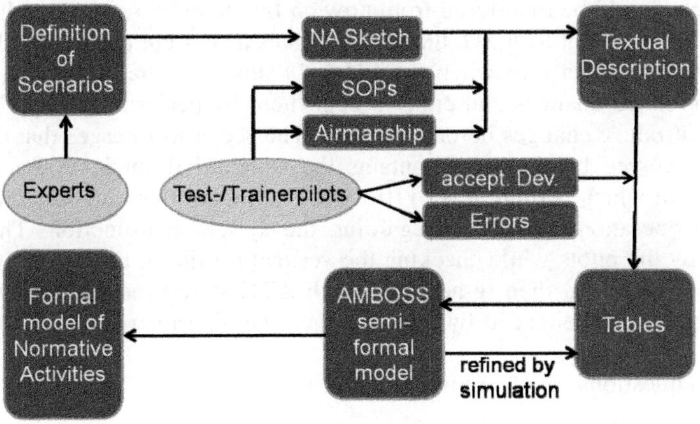

Fig. 2. Modelling Process

The most relevant activities for this paper are those for re-planning. Re-planning means modifying the current flight route via the AHMI by changing the lateral or vertical profile. Changes to the route can be initiated either by the pilots or by the controllers. In the first case the pilots introduce the changes into the route and send it down to ATC (downlink). In the latter case ATC sends a modified route up to the aircraft (uplink). In both cases the last three actions which have to be performed by the pilots are the same: (*NA1*) Generate the modified route by clicking on the "Dirto" button (Fig. 1, bottom left), as a result the new trajectory is shown as a dotted line; (*NA2*) click the "Send to ATC" button (Fig. 1, bottom middle) to downlink self-initiated changes or to acknowledge uplinked changes; (*NA3*) next a feedback from ATC is received in the form of an uplink. Since this uplink may contain further lateral or vertical changes pilots must check the lateral and vertical profile to identify any final modifications. If a change introduced by ATC at this stage is not acceptable for any reason, then the re-planning procedure has to be restarted by the pilots resulting in a new downlink. In case no changes have been received or the changes are acceptable pilots have to press the "Engage" button (Fig. 1, bottom right) to activate the new route.

The trajectory in Fig. 1 represents a typical re-planning scenario which we used in the HUMAN project to generate detailed hypotheses on pilot behavior (see Section 5) provoking Learned Carelessness and Cognitive Lockup. It starts during cruising at flight level 250 (25,000 feet) on a flight inbound to FRA (Frankfurt, Germany).

Passing waypoint ETAGO (approx. 130NM inbound to Frankfurt), a system non-normal message pops up advising the crew of a fuel-pump malfunction. The normative activities require the crew to initiate descent to maximum flight level 100 in order to assure adequate pressure for continuous fuel feed to the respective engine (approx. 60NM earlier than planned). This will be done by a cruise-level alteration in the current flight route via the AHMI followed by a trajectory generation, negotiation and activation (steps NA1, NA2 and NA3) as described above. During descent, the crew receives the latest weather report of Frankfurt which allows preparing for the given approach. The report indicates that there is a thunderstorm approaching the airport which should be monitored from now on by the crew on the weather radar. In the vicinity of waypoint ROLSO, the crew receives a shortcut uplink which clears the flight to proceed directly to waypoint CHA. In this case the pilots are required to check the uplinked changes and either accept them by performing steps NA1, NA2, NA3 or to introduce changes before doing so. The scenario foresees that during NA3 the uplink received by the crew contains the standard flight level for the current arrival segment which is flight level 110, 1000 feet higher than the previous clearance and off the operational envelope regarding the system malfunction. This is to be recognized by the pilots while checking the vertical profile of the uplink, who should correct the altitude and then re-negotiate with ATC starting again with NA1. If the incorrect altitude was engaged by the crew then the aircraft would re-climb to flight level 110.

The main questions which are investigated are:

- Does the pilot model recognize the incorrect altitude?
- Is the pilot model able to recover from the re-climb by initiating a new descent via the AHMI?

In Section 5, we show that the approaching thunderstorm may have a significant effect on the error recovery.

3 Cognitive Processes Involved in Re-planning Tasks

To explain and model why pilots deviate from normative activities, we have focussed on the underlying cognitive processes. In this section, we describe cognitive processes that play a role in re-planning and that are the basis for our crew model.

Cognitive processes can be differentiated by their degree of consciousness. Rasmussen [5] defines three different behaviour levels in which cognitive processing and hence errors can take place: skill-based, rule-based, and knowledge-based behaviour. The level of processing mainly depends on the experience with a task. Anderson [1] distinguishes very similar levels but uses the terminology of autonomous, associative, and cognitive level.

A task that is encountered for the first time is processed on the cognitive level with maximal cognitive effort. This processing is goal driven; alternative plans to reach a goal are evaluated usually through mental simulation, and finally one plan is selected to be executed. With some experience, the associative level is used, where solutions are stored that proved to be successful; the pilot has for example learned how to handle the cockpit systems in specific flight scenarios. According to Rasmussen [5], processing is controlled by a set of rules that have to be retrieved and then executed in the appropriate context. On the autonomous level routine behaviour emerges that is applied without conscious thought, e.g. manually manoeuvring an aircraft.

When solving a task, people tend to apply a solution on the lower levels first, and only revert to solutions on higher levels when lower-level ones are not available [5] or when the situation requires very careful handling due to unusual and safety relevant conditions.

In our research, we focus on two kinds of error production mechanisms that we associate with the associative and the cognitive level respectively, namely Learned Carelessness and Cognitive Lockup.

Learned Carelessness: When re-planning takes place on the associative layer, the procedure may be simplified according to scenarios encountered before. The psychological theory of Learned Carelessness states that humans have a tendency to neglect safety precautions if this has immediate advantages, e.g. it saves time because less physical or cognitive resources are necessary [11]. Careless behaviour emerges if safety precautions have been followed several times but would not have been necessary, because no hazards occurred. Then, people tend to omit the safety precautions and the absence of hazardous consequences acts as a negative reinforcer of careless behaviour.

Cognitive Lockup: On the cognitive layer, the cognitive attention may be captured by a task, which causes people to switch between tasks too late or not at all. This usually happens in situations with a high multitask workload, as switching between tasks costs time and effort, and cognitive resources are limited [3].

4 Modelling Re-planning in a Layered Cognitive Architecture

Cognitive architectures were established in the early eighties as research tools to unify psychological models of particular cognitive processes [12]. These early models only dealt with laboratory tasks in non-dynamic environments [13], [14]. Furthermore, they neglected processes such as multitasking, perception and motor control that are essential for predicting human interaction with complex systems in highly dynamic environments like the air traffic environment addressed in HUMAN with the AFMS target system. Models such as ACT-R and SOAR have been extended in this direction [15], [18] but still have their main focus on processes suitable for static, non-interruptive environments. Other cognitive models like MIDAS [16], APEX [17] and COGNET [19] were explicitly motivated by the needs of human-machine interaction and thus focused for example on multitasking right from the beginning. To our knowledge, none of these architectures has a multi-layered knowledge processing, with different levels of consciousness, as proposed in the following.

4.1 The Cognitive Architecture CASCaS

In HUMAN the cognitive architecture CASCaS (Cognitive Architecture for Safety Critical Task Simulation) as depicted in Fig. 3 is used to model the cognitive process described in the previous section. CASCaS is based on research performed by OFFIS in the European project ISAAC (6th Framework Programme) [8], and has been extended in HUMAN to cover two of Anderson's behaviour levels (c.f. section 3).

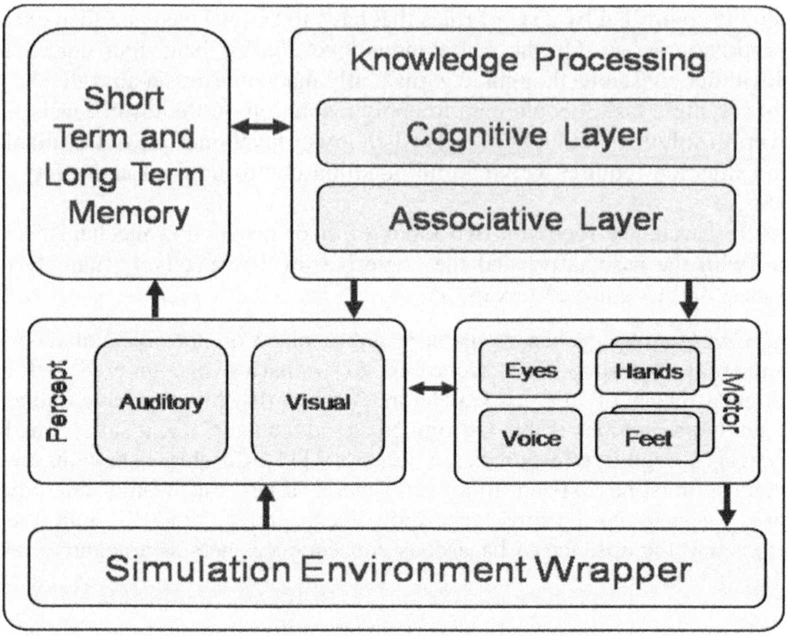

Fig. 3. CASCaS Architecture

The core of CASCaS is formed by the layered *knowledge processing* component that contains the associative and the cognitive layer.

Knowledge for both layers is stored in the *memory component*. The *short-term memory* stores variable-value pairs of data that have been perceived from the environment or derived by applying rules (see below). The *long-term memory* stores flight procedures in form of Goal-State-Means (GSM) rules (Fig. 3). All rules consist of a left-hand side and a right-hand side. The left-hand side consists of a goal in the Goal-Part and a State-part, which specifies Boolean conditions on the current state of the environment, together with associated memory-read items to specify variables that have to be retrieved from memory. The right-hand side consists of a Means-Part containing motor as well as percept actions (e.g. hand movements or attention shifts), memory-store items and a set of partial ordered sub-goals.

Rule 1 in Fig. 4 defines a goal-sub-goal relation between HANDLE_ATC_ UPLINK and the three sub-goals GENERATE_ROUTE, NEGOTIATE_ROUTE and CHECK_ATC_UPLINK_VERT_PREPARE. The precondition in the goal term imposes a temporal order on the sub-goals, i.e. NEGOTIATE_ROUTE can only be performed after GENERATE_ROUTE.

Additionally to the GSM-rules we added a second rule type, called reactive rules. Rule 2 in Fig. 4 is an example for this rule type. The only difference is that reactive rules have no Goal-Part. While GSM-rules represent deliberate behaviour, and are selected by the knowledge processing component during the execution of a flight procedure, reactive rules (State-Means rules) represent immediate or reactive behaviour which is triggered by events in the environment, e.g. in rule 2 of Fig. 4 an ATC uplink message (atc_uplink_mesage==true) triggers the goal HANDLE_ATC_ UPLINK.

	Rule 1:	
IF	Goal (HANDLE_ATC_UPLINK)	(G)oal-Part
IF	Memory-Read(atc_uplink_present)	(S)tate-Part
IF	Condition(atc_uplink_present==true)	
	→	
THEN	Goal (GENERATE_ROUTE)	(M)ean-Part
THEN	Goal (NEGOTIATE_ROUTE, precondition=GENERATE_ROUTE)	
THEN	Goal (CHECK_ATC_UPLINK_VERT_PREPARE, precondition=NEGOTIATE_ROUTE)	

	Rule 2:	
IF	Condition(atc_uplink_message==true)	(S)tate-Part
	→	
THEN	Memory-Store (atc_uplink_present, true)	(M)ean-Part
	Goal (HANDLE_ATC_UPLINK)	

Fig. 4. Format of GSM rules

The *associative layer* selects and executes rules from long-term memory. It is modelled as a production system. Characteristic for such systems is a serial cognitive cycle for processing rules: A goal is selected from the set of active goals (Phase 1), all rules containing the selected goal in their Goal-Part are collected and a short-term memory retrieval of all state variables in the Boolean conditions of the collected rules is performed (Phase 2). If a variable is absent in memory, a dedicated percept action is fired and sent to the percept component to perceive the value from the environment and to write it into the short-term memory. After all variables have been retrieved, one of the collected rules is selected by evaluating the conditions (Phase 3). Finally the selected rule is fired (Phase 4), which means that the motor and percept actions are sent to the motor and percept component respectively and the sub-goals are added to the set of active goals. This cycle is started when a Boolean condition of a reactive rule is true. In Phase 2 reactive rules may be added to the set of collected rules if new values for the variables contained in the State-Part have been added to the memory component (by the percept component). In Phase 3, reactive rules are always preferred to non-reactive rules. The cognitive cycle is iterated until no more rules are applicable.

The *cognitive layer* reasons about the current situation and makes decisions based on this reasoning. Consequently, we differentiate between a decision-making module, a module for task execution and a module for interpreting perceived knowledge (sign-symbol translator).

The *decision-making module* determines which goal is executed. Goals have priorities, which depend on several factors: first, goals have a static priority value that is set by a domain expert. Second, priorities of goals increase over time if not executed. Implicitly, temporal deadlines are modelled in this way. If, while executing a goal, another goal has a clearly higher priority than the current one, the execution of the current goal is stopped and the new goal is attended to.

The *task-execution module* executes the goals that have been chosen by the decision-making module. (Sub-)tasks might be passed to the associative layer if rules exist in long-term memory.

The *sign-symbol translator* is based on Rasmussen's differentiation between signs and symbols [5]. This module raises the level of abstraction of the signs perceived by the percept component and stored in short-term memory by identifying and interpreting the situation, and thereby adding extra knowledge to the sign. In addition, background knowledge is applied to judge and evaluate the current situation.

The associative and cognitive layer interact in the following ways: first, the cognitive layer can start (and thus delegate), monitor, temporally halt, resume and stop activities on the associative layer by manipulating the associative layer's goal agenda. Monitoring of the associative layer is realized through determining whether the appropriate goals are placed in the goal agenda.

The associative layer can inform the cognitive layer about the status of rule execution, e.g. current execution is stuck because for the chosen goal no rules are available in long-term memory or execution of a perceived event cannot be started for the very same reason. In these cases the cognitive layer starts to perform the goal or event. Furthermore, the cognitive layer can take over control at any time. Currently this is initiated by setting the parameter "Consciousness". If the value is "associative" then every event will first be processed if possible and the cognitive layer becomes only active if no rules are available. If the value is "cognitive" then the cognitive layer processes each event independent of the availability of rules.

The *percept* component consists of two sub-components, an *auditory* component for receiving sounds or vocal input (in form of variables representing acoustic input), and a *visual* component for perception of visual input (in form of variables representing visual input). While the auditory component is purely reactive to external input, the visual component can be controlled by the knowledge processing component via percept-actions contained in rules. Percept-actions result in eye movements, which are performed by the *eyes* sub-component in the motor component. The eyes component has a detailed model of eye movements, in order to simulate the timing of the eye movements. For a more detailed description of the visual component, see [4]. All information that has been perceived is stored in the short-term memory of the cognitive architecture.

The *motor* component contains, apart from the eye component, modules for *hand* and *feet* movement. This components use the 2D and 3D formulations of Fitt's Law [10] in order to model the timing of the requested movements (via motor actions received from the knowledge processing component). With these components, the cognitive model can for example simulate button presses.

The *Simulation Environment Wrapper* provides data for the percept component and functions for the motor component to manipulate the environment by connecting CASCaS with different simulation backends. In HUMAN we connected CASCaS to the fixed base flight simulator used by the DLR for experiments with human pilots. In this way the model can be executed and data can be recorded in the very same environment in which also human subject pilots interact. This allows validation of the model by comparing model data with human data.

4.2 The Error-Producing Mechanisms

Learned Carelessness is modelled on the associative layer by a dedicated learning algorithm. This is modelled by melting two rules into one rule by means of *rule composition* [8]. A precondition for composing rules is that firing of the first rule has evoked the second rule, or more exact, the first rule derives a sub-goal that is contained in the Goal-Part of the second rule. Melting the rules means building a composite rule by combing the left-hand sides of both rules and also combing both right-hand sides. The crucial point is that in this process elements that are contained on the right-hand side of the first and also on the left hand side of the second rule are eliminated. This process cuts off intermediate knowledge processing steps.

Rule 5 in Fig. 5 specifies that it is only allowed to proceed with engaging the route [Goal(ENGAGE_ROUTE)] if the vertical profile contains no changes (changes_present == false). Using rule 3 the current value of the variable is perceived from the AHMI. Rule 4 stores the perceived value into the short-term memory. Mostly when pilots want to engage a route, there are actually no changes to the vertical profile. Thus, most of the time the percept action delivers 'false'. Our pilot model produces a new simplified rule by merging rule 3 and 4 to rule 71, where the existence of changes is no longer perceived from the AHMI but is just retrieved from memory. The percept action has been eliminated and the simplified rule always stores the value 'false' into the memory. Applying rule 71 results in careless behaviour: engaging an uplinked route independent from actual changes in the vertical profile. At the beginning of the simulation, all rules in the long-term memory component are normative, meaning that the application of these rules does not lead to an error.

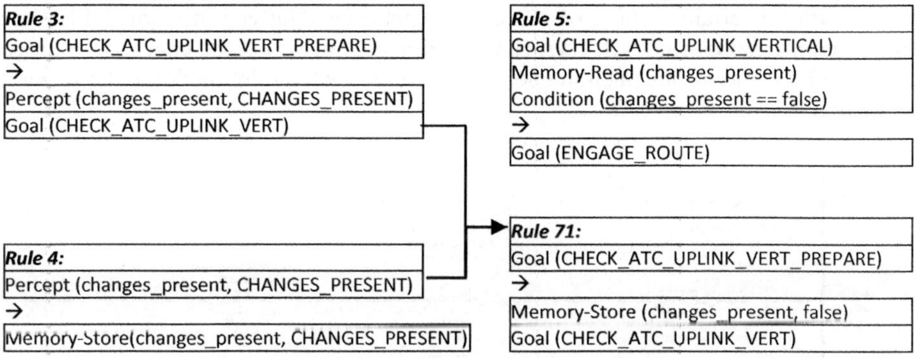

Fig. 5. Composition of rule 3 and 4 to rule 71, which lead in rule 5 to careless behaviour

Cognitive lockup is implemented as part of the goal decision mechanism, thus on the cognitive layer. In certain situations, switching between goals does not take place even though the priority of another goal is higher than the currently selected one. The selection mechanism is extended by the parameter Task Switch Costs (TSC), which determines the difference that the priorities need to have to halt the execution of a goal to select a different goal to be executed. Task Switch Costs are described extensively in the literature (e.g. [9]). The TSC depends on the cognitive demands of the current task. The higher the cognitive demands the higher are the costs to switch a task: TSC = StartTSC + cognitive_complexity_current_task. The parameter StartTSC denotes the threshold that gives the difference in priority two goals need to have to make an interruption of the one goal and the changing to the other goal possible. This parameter is determined by experimentation. The cognitive complexity of a task is determined by a domain expert and increases the threshold to switch tasks.

5 Detailed Hypotheses on Re-planning Behaviour

This section describes hypotheses on pilot behaviour that have been derived by executing the cognitive crew model in the flight scenario described in Section 2. The hypotheses will in the future be used to validate the model behaviour by comparing the simulation traces of the model with traces of real pilot behaviour. Our hypotheses describe predictions generated by the model with regard to a pilot error due to Learned Carelessness, a pilot error due to Cognitive Lockup and the interaction between both mechanisms in the course of error recovery. The predictions are presented in the form of simulation traces.

Hypothesis 1: *If checking the vertical profile never shows any irregularities, Learned Carelessness will inhibit this check in the future*

The re-planning procedure prescribes to check the vertical profile after the acknowledge has been received from ATC. It can happen that ATC does not accept the altitude that has been downlinked via the AHMI. In this case altitude changes can be seen in the vertical profile. Since this check costs effort, in terms of time needed for goal selection, percept and motor actions, and since altitude changes by ATC are rather unlikely in that phase of the re-planning procedure the check is prone to be omitted after a certain number of procedure repetitions. Our cognitive model learns a simplified procedure rule (c.f. rule 71 in Section 4.2) in which the check is no longer present. Fig. 6 shows this phenomenon as generated by the pilot model in the scenario of Section 2.

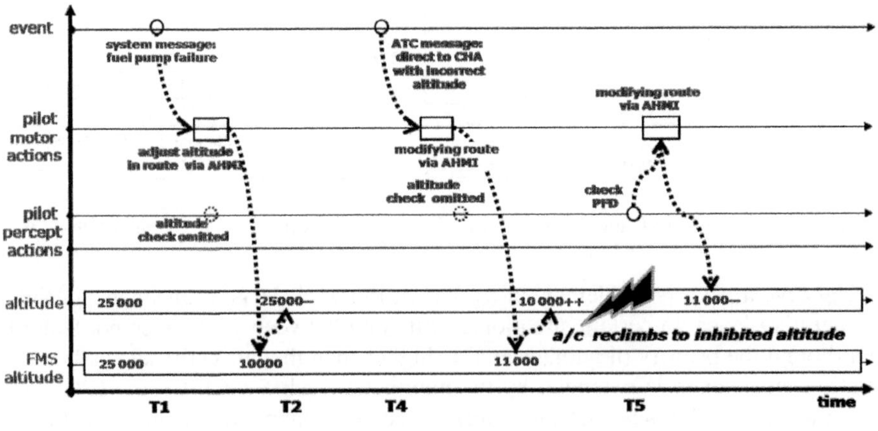

Fig. 6. Pilot error due to Learned Carelessness

At the beginning of the scenario the model has already flown two other experimental scenarios with twelve re-planning events. A simplified rule without the vertical profile check has been learned in our simulations after the 10th procedure repetition and was first applied during the 11th repetition. At T1 in scenario C the fuel pump fails which requires the pilots to descend to altitude 10000. The pilot model adjusts the altitude of the current route via the AHMI, sends it to ATC and receives an acknowledge which is then engaged. The altitude is not checked but in this case there are no consequences. At T4 ATC sends a shortcut allowing the aircraft to fly directly to waypoint CHA. This uplink contains a vertical profile (altitude 11000) that violates the altitude constraint (altitude 10000) which still holds due to the fuel pump malfunction. The model does not notice this violation because it again omits the altitude check before engaging the changed route. Thus, the aircraft starts to re-climb to altitude 11000. After a certain while, at T5, the model recognizes the climb during regular monitoring of the flight conditions. The model corrects the vertical profile of the route via the AHMI which make the aircraft descend again.

Hypothesis 2: *If the cognitive layer keeps control of the proceduralized check on the associative layer, irregularities in the vertical profile will be detected*

We built an alternative version of the pilot model in which the parameter *Consciousness* is set to "cognitive" whenever a system failure is experienced. This value is maintained until the problem is solved. This version of the model has been used to derive an alternative hypothesis for the same scenario (Fig. 7). At T1 the fuel pump failure occurs and *Consciousness* is set to "cognitive". As a consequence the pilot model performs the modification of the route after T4 on the cognitive layer and thus the original non-careless version of the re-planning procedure is applied. The pilot model recognizes the incorrect altitude, corrects it and sends it to ATC, where the change is accepted and sent back.

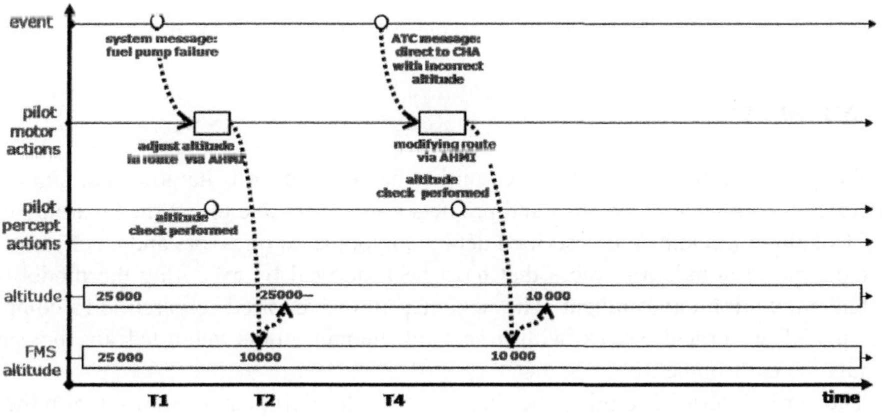

Fig. 7. Conscious procedure execution prevents pilot error

Hypothesis 3: *When a task requires high cognitive demand, other tasks might be inadequately neglected – this Cognitive Lockup will delay subsequent recoveries of irregularities in the vertical profile that were not detected by the associative layer due to Learned Carelessness.*

For this hypothesis we assume a variant of the scenario with an additional event which is emitted at T3. Pilots receive a weather report update indicating that there is a thunderstorm approaching the airport which should be monitored from now on by the crew on the weather radar. As a result the pilot model is focused on monitoring the thunderstorm that the climb of the aircraft due to the incorrect uplink at T4 is recognized considerably later than in the preceding scenario. The reason is the Cognitive Lockup mechanism. The model does not switch to the regular task of monitoring the flight conditions because monitoring of the thunderstorm is a demanding task.

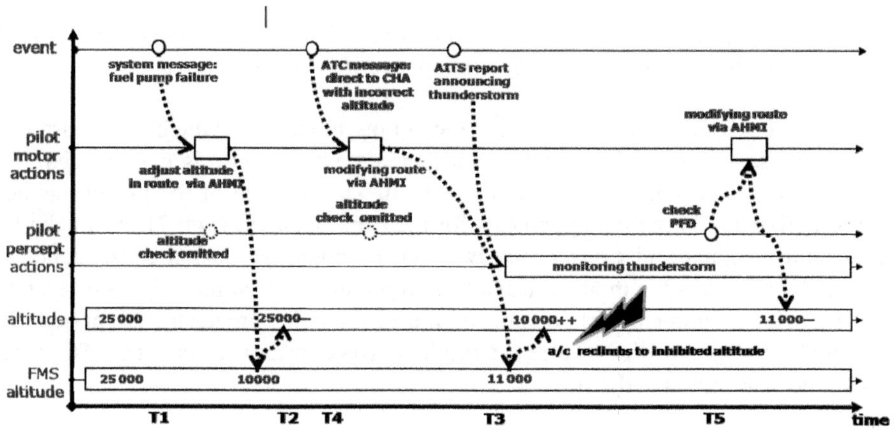

Fig. 8. Error is not recovered due to Cognitive Lockup

6 Summary

In this paper we have presented a cognitive model of pilot behaviour that simulates interaction with cockpit systems and predicts pilot errors due to Learned Carelessness and Cognitive Lockup. We described detailed hypotheses on errors and error recovery in the form of simulation traces that have been derived by executing the model in a virtual simulation environment. The next step in the reported research is to compare the model generated traces with traces of human pilots recorded in the same simulation environment.

The work described in this paper is funded by the European Commission in the 7th Framework Programme, Transportation under the number FP7 – 211988.

References

1. Anderson, J.R.: Learning and Memory. John Wiley & Sons, Inc., Chichester (2000)
2. Edwards, E.: The Emergence of Aviation Ergonomics. In: Wiener, Nagel (eds.) Human Factors in Aviation. Academic Press, San Diego (1988)
3. Kerstholt, J.H.: Dynamic Decision Making. Universiteit of Amsterdam (1996)

4. Osterloh, J.-P., Lüdtke, A.: Analyzing the Ergonomics of Aircraft Cockpits Using Cognitive Models. In: Karowski, W., Salvendy, G. (Hrsg.) Proceedings of the 2nd International Conference on Applied Human Factors and Ergonomic (AHFE), July 14-17. USA Publishing, Las Vegas (2008)
5. Rasmussen, J.: Skills, Rules, Knowledge: Signals, Signs and Symbols and other Distinctions in Human Performance Models. IEEE Transactions: Systems, Man and Cybernetics, SMC-13, 257–267 (1983)
6. Sherry, L., Polson, P., Feary, M., Palmer, E.: When Does the MCDU Interface Work Well? In: International Conference on HCI-Aerro, Cambridge, MA (2002)
7. Dekker, S.: Failure to Adapt or Adaptions that fail. Applied Ergonomics 34(3), 233–238 (2003)
8. Lüdtke, A., Cavallo, A., Christophe, L., Cifaldi, M., Fabbri, M., Javaux, D.: Human Error Analysis based on a Cognitive Architecture. In: Reuzeau, F., Corker, K., Boy, G. (eds.) Proceedings of HCI-Aero, Cépaduès-Editions, France, pp. 40–47 (2006)
9. Liefooghe, B., Barrouillet, P., Vandierendonck, A., Camos, V.: Working Memory Costs of Task Switching. Journal of Experimental Psychology: Learning, Memory, & Cognition 34, 478–494 (2008)
10. Grossmann, T., Balakrishnan, R.: Pointing at Trivariate Targets in 3D Environments. In: CHI 2004: Proceedings of SIGCHI, pp. 447–545. ACM Press, New York (2004)
11. Frey, D., Schulz-Hardt, S.: Eine Theorie der Gelernten Sorglosigkeit. In: Mandl, H. (Hrsg.) 40. Kongress der Deutschen Gesellschaft für Psychologie, pp. 604–611. Hogrefe Verlag für Psychologie, Göttingen (1997)
12. Newell, A.: Unified Theories of Cognition. Harvard University Press (1994); Reprint edition
13. Anderson, J.R.: Rules of Mind. Lawrence Erlbaum Associates, Hillsdale (1993)
14. Newell, A., Rosenbloom, P.S., Laird, J.E.: Symbolic Architectures for Cognition. In: Posner, M.I. (ed.) Foundations of Cognitive Science, pp. 93–131. MIT Press, Cambridge (1989)
15. Anderson, J.R., Bothell, D., Byrne, M.D., Douglass, S., Lebiere, C., Qin, Y.: An integrated theory of the mind. Psychological Review 111(4), 1036–1060 (2004)
16. Corker, K.M.: Cognitive models and control: Human and system dynamics in advanced airspace operations. In: Sarter, N., Amalberti, R. (eds.) Cognitive Engineering in the Aviation Domain, pp. 13–42. Lawrence Erlbaum Associates, Mahwah (2000)
17. Freed, M.: Simulating Human Performance in Complex, Dynamic Environments. PhD thesis, Northwestern University (1998)
18. Wray, R., Jones, R.: An introduction to Soar as an agent architecture. In: Sun, R. (ed.) Cognition and Multi-agent Interaction: From Cognitive Modeling to Social Simulation, pp. 53–78. Cambridge University Press, Cambridge (2005)
19. Zachary, W., Santarelli, T., Ryder, J., Stokes, J., Scolaro, D.: Developing a multi-tasking cognitive agent using the COGNET/iGEN integrative architecture. In: Proceedings of 10th Conference on Computer Generated Forces and Behavioral Representation, pp. 79–90. Simulation Interoperability Standards Organization, Norfolk (2001)
20. Frische, F., Mistrzyk, T., Lüdtke, A.: Detection of Pilot Errors in Data by Combining Task Modeling and Model Checking. In: Gross, T., Gulliksen, J., Kotzé, P., Oestreicher, L., Palanque, P., Prates, R.O., Winckler, M. (eds.) INTERACT 2009, Part I. LNCS, vol. 5726, pp. 528–531. Springer, Heidelberg (2009)

The Perseveration Syndrome in the Pilot's Activity: Guidelines and Cognitive Countermeasures

Frédéric Dehais[1], Catherine Tessier[2], Laure Christophe[3], and Florence Reuzeau[3]

[1] Université de Toulouse, ISAE, Centre Aéronautique et Spatial, 10 av. E. Belin, 31055 Toulouse Cedex 4, France
frederic.dehais@isae.fr
[2] Onera DCSD, 2 av. E. Belin, 31055 Toulouse Cedex 4, France
catherine.tessier@onera.fr
[3] Airbus France, 316 route de Bayonne, 31060 Toulouse Cedex 9, France
{laure.christophe,florence.reuzeau}@airbus.com

Abstract. In this paper we present the Ghost project, an Airbus research program that aims at preventing aircrews from the perseveration syndrome. This particular behavior is known to summon up all the pilots' mental efforts toward a unique objective even if the latter is dangerous in terms of safety. The unification of cognitive psychology and neuropsychology theories tends to prove that such a behavior comes from an impairment in attention shifting mechanisms induced by stressing situations. Such an approach paves the way to design cognitive countermeasures dedicated to enhance the pilot's attention shifting capabilities. Two preliminary experiments are presented to test these hypotheses and concepts.

Keywords: Perseveration syndrome, cognitive countermeasures, experiments.

1 Introduction

Since 2004 Airbus France has been funding a 4-year research program called the "Ghost project" in cooperation with Isae (Institut Superieur de l'Aéronautique et de l'Espace) and Onera (Office National d'Etudes et de Recherches Aérospatiales). This research program deals with the perseveration syndrome in civilian aviation. This particular behavior [1] is known to summon up all the pilot's mental efforts toward a unique objective (excessive focus on a single item of a display or excessive focus of the pilot's reasoning on a single task). Once entangled in perseveration, the pilots do everything to succeed in their objective (ex: landing) even if it is dangerous in terms of safety. Worst, their reasoning capabilities suffers from the confirmation bias [2] that leads them to neglect any environmental clues that could question their reasoning (i.e. audio and visual alarms). Social capacities are also degraded with aggressiveness or loss of communication within the aircrew. Therefore, the Ghost project aims at:

- understanding the underlying mechanisms of perseveration;
- identifying situations that create perseveration and their precursors in civilian aviation;

The original version of this chapter was revised: The copyright line was incorrect. This has been corrected. The Erratum to this chapter is available at DOI: 10.1007/978-3-642-11750-3_10

P. Palanque, J. Vanderdonckt, and M. Winckler (Eds.): HESSD 2009, LNCS 5962, pp. 68–80, 2010.

- designing cognitive countermeasures to help the pilots to recover from these situations;
- developing an experimental environment and "perseverogenic" scenarios to test the different hypotheses and the cognitive countermeasures;
- implementing formal tools [3] to automatically detect perseveration behaviors and send accurate cognitive countermeasures thanks to on-line flight parameters analyses.

In this paper, we firstly discuss some possible explanations of the underlying mechanisms of perseveration based on psychosociology, cognitive psychology, and neurosciences theories and accident analyses. Such an approach paves the way to propose guidelines to design cognitive countermeasures based on a neuroergonomics approach [4, 5, 6]. Then, the very first experiments of the Ghost project are presented. Eventually the final discussion focuses on the Airbus methodology to design cognitive countermeasures.

2 Perseveration: Guidelines

2.1 Related Work

Classical psychosociology theories [7, 8, 9] demonstrate that the higher and longer the level of commitment to achieve a goal the harder it is to drop the goal even if it is not relevant anymore. This can be illustrated by an accident (MD82, Little Rock, 1999) where an aircrew after a very long flight has persevered in a wrong landing decision despite a particular combination of bad conditions at the final destination (thunderstorm, heavy rain, strong cross wind and then windshear). Unfortunately, this case is not unique, and a study conducted by the MIT [10] demonstrates that in 2000 cases of approaches under thunderstorm conditions, two aircrews out of three keep on landing especially if their flight has been delayed, or if they follow another airplane or if it is a night flight. These psycho-social theories may also provide explanations to a recently published report of the BEA (the French national institute for air accident analysis) revealing that pilots' erroneous perseveration behaviors (the "get-home-itis syndrome") have been responsible for more than 41,5 percent of casualties in light aircrafts [11]. Eventually an interesting report written by the French military safety board [12] shows that if the loss of control of a fighter aircraft is the consequence of the pilot's fault then the decision to eject the aircraft is much harder to take for the pilot than if the loss of control is independent of his will (ex: system failure): most of the time, the pilot's objective is to stay in his cockpit to fix his error, being unaware of the proximity of the ground.

Research in cognitive psychology also brings interesting contributions to understand perseveration. As mentioned in the first section, such a behavior may impair decision making capabilities with a trend to confirmation bias [2] as long as the pilots search for or interpret new information in a way that confirms their preconceptions, which makes them irrationally avoid any decision that would contradict their beliefs. In the Little Rock accident and despite a heavy rain and a strong crosswind, the captain faced such a confirmation bias, as he "voluntary" confounded dry runway crosswind limitation with wet runway crosswind limitation, which led him to keep on landing. Eventually in their last try to land, dangerous windshear alerts did not even incite them to go around, but to choose another more favorable runway! Many experiments have addressed this difficulty for pilots to revise their flight plan [13, 14, 15, 16] in particular during the stressing landing flight phase [17].

Another associated concept is known as fixation errors. The typology shows that parallels can be drawn between fixation errors and perseveration [18]:

- type 1: the human operator is unable to make his mind to achieve his current goal;
- type 2: the human operator keeps on doing the sequence without any control;
- type 3: the operator has overconfidence in his strategy and neglects or does not trust any external data (ex : alarms).

Hypovigilance, high level of stress, high workload are suspected to be the causes of this attentional impairment.

One may say that this approach is essentially descriptive and does not provide explanations of the underlying mechanisms, in particular as long as fuzzy concepts like workload are not measurable. Nevertheless it is worth noticing that the same kinds of erroneous behaviors are observed both in stressed operators (e.g.: a pilot facing a major breakdown) and in brain-injured patients (i.e. dysexecutive syndrome) performing a complex cognitive task: in this sense Pastor [19] suggests the existence of a cognitive continuum between pathology and normal but stressed persons. Experiments conducted with these two populations of subjects in a microworld environment show trends to perseveration and fixation behaviors for all subjects. Interestingly enough, a review of the dysexecutive syndrome shows strong similarities with the fixation error typology [18, 20]:

- abulia: the patient is unable to make up his mind for simple choices;
- perseveration on psychomotor response;
- incapacity to adapt to environmental changes.

Research in neuropsychology suggests that these behaviors result from an incapacity to inhibit a current task to shift to another one: the perseveration syndrome is an inhibition impairment that may be caused by the irremediable loss of an associated neural network (i.e. cerebral lesion) [21] or a temporary loss of a cognitive function induced by stress and emotions [22, 23].

Therefore we propose to define the perseveration syndrome as an incapacity for an agent to shift from a goal to a new one in order to react adequately to the evolution of the environment.

2.2 Cognitive Countermeasures

One must consider that the perseveration syndrome leads the user interface designers to face a paradox: how can it be expected to "cure" the pilots of perseveration if the alarms/systems designed to warn them are not perceived? Therefore an objective is to design cognitive countermeasures that we define as a means to mitigate a cognitive bias [1]. As mentioned in the previous section, an interesting approach is to consider the recent explanatory models from neurosciences and neuropsychology to find out solutions to this paradox. In this sense Posner [24] suggests that the attentional processes are based on a three-step mechanism: shifting, orienting, focusing. This means that to gaze at an item, one must first shift his/her attention to the current item, to reorient and then to focus on the new selected item. These mechanisms have been described thanks to the study of brain-injured patients with specific attentional impairment. In addition, Camus [25] postulates the existence of a functional coupling between a posterior system dedicated to gaze orientation and an anterior (prefrontal) system responsible for the voluntary control of attention. Therefore considering the paradigm of a continuum between brain-injured patients and stressed operators [18], it then may be assumed that operators who face a perseveration syndrome suffer from an attention shifting impairment.

This gives us clues to understand the pilot's not noticing audio nor visual alarms while persevering: the latter are based on information adding for orienting and focusing attention (e.g. by prioritizing the level of alarms), but not for shifting attention.

The idea is then to let the user interface proceeds attention shifting by providing an accurate stimulus into his visual field. Therefore the principle of cognitive countermeasures relies on:

- a subtle modification of information presentation on which the pilot focuses;
- its replacement by an accurate visual stimulus to change the pilot's focus.

In order to test this hypothesis, two exploratory experiments are described in the next sections.

3 Preliminary Experiments

3.1 Material and Methods

3.1.1 Participants and Flight Simulator

21 pilots participated in the experiments. The pilots' flying experience ranged from novice (5 hours, small aircraft) to very experienced (3500 hours, military aircraft). A PC-based flight simulator (Cessna 310 flight model – Flight Gear simulator) equipped with a stick, a rudder and a thrust is used for the purpose of the experiment. A wizard of Oz is also implemented in order to let a human operator trigger the cognitive countermeasures.

3.1.2 Scenarios and Procedure

For the purpose of the experiments, three "perseverogenic" scenarios are designed. All of them rely on the principle of a progressive weather change close to the final goal (i.e. landing ground): as long as it is difficult for the pilots to detect a progressive weather change [13, 14, 15, 16] especially when close to final destination [11], it is expected to provoke a "perseveration" syndrome.

- scenario 1 is a visual flight rules (VFR) navigation task from Toulouse-Blagnac airport to Francazal airport including three waypoints (VOR 117.70, NDB 331 and NDB 423). The visibility is slowly degraded then progressively enhanced in a such way that the pilot cannot see the runway but at the last moment when it is too late to land on it.

- scenario 2 is a VFR navigation task from Toulouse-Blagnac airport to Francazal airport including three waypoints (VOR 117.70, NDB 415 and NDB 423). The visibility is decreased from the runway threshold: from far away, the runway is visible but as the pilot gets closer to it, it disappears progressively in a thick fog that eventually prevents him from landing.
- scenario 3 is a VFR navigation task from Toulouse-Blagnac to Toulouse-Blagnac including three waypoints (VOR 117.70, NDB 331 and NDB 423). The visibility is progressively degraded and as the pilot flies over waypoint 2 (NDB 331) the left engine fails.

Each session lasts about one hour and a half:

- the pilots have a one-hour training period on the PC to learn the maneuverability and the limits of the flight simulator ;
- once the pilots are familiarized with the flight simulator, one of the three scenarios is presented to them (duration : 30 minutes). They are told to perform the flight in VFR and to behave as close as possible as they would in real conditions (*i.e.* to abort the mission in case of weather deterioration).

3.2 Cognitive Countermeasures

In all three scenarios, it is expected that the deterioration of the weather conditions will focus the pilot's attention on an instrument called the H.S.I. (Horizontal Situation Indicator) if he intends to go on with landing. Therefore the countermeasures consist in flashing the H.S.I. during a few seconds and displaying instead a short message that is adapted to the scenario : in scenarios 1 and 2, the HSI is replaced with message "Go back to Blagnac" to interfere with the pilot's initial decision to land at Francazal in poor weather conditions; in scenario 3, the HSI is flashed and replaced with message "Engine failure" to warn the pilot and suggest an emergency landing.

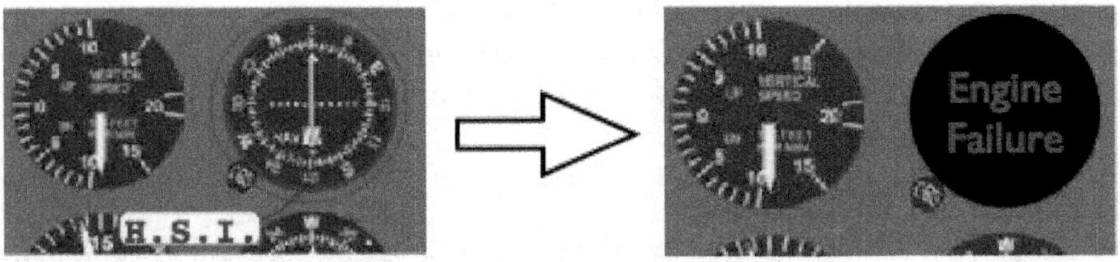

Fig. 1. A cognitive countermeasure for scenario 3: the HSI (circled in red) is flashed and then replaced by message "Engine Failure"

3.3 Experimental Scenarios

3.3.1 Results for Scenarios 1 and 2: "Impossible Landing"

In these scenarios, the pilots faced the decision of mission abortion, i.e. not landing at Francazal and flying back to Blagnac. Both scenarios were tested within two different contexts: without countermeasures and with countermeasures.

• **Condition 1: no countermeasure**. Seven pilots tested scenario 1 or 2 without any countermeasure (see next table). "Circuits" corresponds to the number of circuits performed by the pilots round Francazal airfield before crashing or landing.

Pilot	Scenario	Circuits	Results
Pilot1	1	3	crash
Pilot2	1	3	crash
Pilot3	1	1	chance landing
Pilot4	1	1	chance landing
Pilot5	2	1	crash
Pilot6	2	1	chance landing
Pilot7	2	1	crash

The results suggest that without any countermeasure, none of the pilots came to the right decision (fly back to Blagnac) : they all persevered at trying to land at Francazal.

Four of them hit the ground, and the other three had a "chance landing", which means that while they were flying round Francazal, the runway appeared between two fog banks and they succeeded in achieving a quick landing. During the debriefing, all of them admitted they had come to an erroneous and dangerous decision.

• **Condition 2: with countermeasures.** 12 pilots tested scenario 1 or 2 with countermeasures (see next table). "Circuits" corresponds to the number of circuits performed by the pilots round Francazal before a countermeasure was triggered by the wizard of Oz.

Pilot	Scenario	Circuits	Results
Pilot8	1	3	crash on Francazal
Pilot9	1	2	back to Blagnac
Pilot10	1	2	back to Blagnac
Pilot11	1	2	back to Blagnac
Pilot12	2	3	back to Blagnac
Pilot13	2	2	back to Blagnac
Pilot14	2	2	back to Blagnac
Pilot15	2	2	back to Blagnac
Pilot16	1	2	chance landing Francazal
Pilot17	1	2	back to Blagnac
Pilot18	1	2	back to Blagnac
Pilot19	1	5	crash on Francazal

The results show the efficiency of the countermeasures to cure the pilots of perseveration: 9 pilots out of 12 changed their minds thanks to the countermeasures, and flew back safely to Blagnac. During the debriefing, all the pilots confirmed that the countermeasures were immediately responsible for their changing their minds. Moreover the short disappearance of the data due to the countermeasures did not cause any stress to them. 4 military pilots found that the solutions proposed by the countermeasures were close to what a human co-pilot would have proposed. The results with Pilot19 suggest that the more a pilot perseveres, the more difficult it is to get him out of perseveration. During the debriefing, Pilot19 told us that he was obsessed by the idea of landing, and that he became more and more stressed as he was flying round Francazal. He then declared that he did not notice any countermeasures.

Pilot16 and Pilot8 also persevered despite the countermeasures: they declared they knew they were not flying properly and that they had done that on purpose because they wanted to test the flight simulator.

3.3.2 Results for Scenario 3: "Failure"

Only 2 pilots tested scenario 3. One pilot experimented this scenario without any countermeasures: he did not notice the failure and hit the ground. The second pilot was warned through the countermeasures that he had a failure: the pilot immediately performed an emergency landing on the closest landing ground.

3.3.3 Other Results

During the experiments, 9 pilots made errors, e.g. selection of a wrong radio frequency, erroneous altitude, omission to retract flaps and landing gear. To warn them, the wizard of Oz blinked the display on which they were focusing and displayed the error (e.g.: "gear still down"). In each case, the countermeasure was successful: the pilots performed the correct action at once. During the debriefing, the pilots declared that this kind of alarm was very interesting, and much more efficient and stressless than a classical audio or visual alarm because they could identify at once what the problem was.

4 Experiments Conducted with Airbus

The objective is to conduct experiments in a more realistic environment (ISAE 3-axis flight simulator) with Airbus pilots to assess the principle of the cognitive countermeasures and their possible interests for commercial aviation.

4.1 Material and Methods

4.1.1 Participants, Flight Simulator

Three male airline right-handed pilots participated in this experiment, their mean age was 45, and their mean flying experience was 11200 hours. The ISAE-SUPAERO 3-axis motion flight simulator was used to conduct the experiments (see figure 1). It simulates an Airbus 300 flight model, and its user interface is composed of a PFD (Primary Flight Display), an ND (Navigation Display) and the upper ECAM (Electronic Central Aircraft Monitoring Display) page. The left-seated pilot has a fly-by-wire stick to control the flight, a rudder, and two manual thrust levers.

A software is implemented to manage the different events that occur during the landings (see next section). It allows the special events to be triggered e.g. a windshear, an antiskid failure, and the cognitive countermeasures.

4.1.2 Experimental Scenario

The flight scenario is designed to:

- make the pilots persevere in a wrong landing decision;
- make the pilots not notice an alarm (here, an antiskid failure).

Fig. 2. The ISAE flight simulator used for the experimentation

To reach these objectives, the pilots are asked to make 10 night manual landings on the 15L runway at Toulouse Blagnac thanks to the ILS (Instrument Landing System). These manual landings are "Cat II landings", which means that the touch down zone must be visualized before a height of 100 ft. Each new landing is more difficult than the previous one as landing conditions slightly change from one another (i.e.: stronger crosswind, lower visibility…) During the tenth and last landing, the pilots face a slight windshear close to the ground (150 feet QFE) under rainy conditions. In addition an antiskid failure is also triggered consisting of an audio alarm (master caution – "single chime") and an associated visual message on the ECAM. Such a combination of degraded conditions should make the pilots go-around but we hypothesize that:

- this repetitive landing task will alter the pilot's level of alertness and decision making capacities;
- the progressively degrading flight conditions will tend to focus the pilots' attention on their landing path and their airspeed on the PFD to the detriment of the monitoring of the other flight parameters (i.e. alarms);
- this combination of factors will make the subjects not go-around and also neglect the antiskid alarm during the last landing.

4.1.3 Cognitive Countermeasures

The cognitive countermeasures are designed to break the mechanism of perseveration in order to make the pilots understand that their last approach is not stabilized and to make them aware of the failure. The design of the countermeasures relies on the principle proposed in the previous section.

4.1.4 Procedure

As only three pilots participated in this experiment, each experimental session was conducted with one pilot at once (*i.e.* without the presence of a copilot). Each session lasted about one hour and a half:

- the pilots had a 20-minute training period on the flight simulator to learn the maneuverability and the limits of the flight simulator;
- once the pilots were familiarized with the flight simulator, the scenario was presented to them and they were asked to perform the 10 successive landings with respect to the flight safety (ex : a stabilized approach, touch down zone visualized before 100 ft...) So it was clearly explained that any missed approach had to lead to a go-around. Each landing lasted about 2 minutes and a half.

The pilots were not told about the real purpose of the experiment (i.e. the cognitive countermeasures): they were told that the objective of the experiment was to assess a tool to detect non-stabilized approaches.

Pilot 1 and Pilot 2 were the tested subjects as cognitive countermeasures were sent to them; Pilot 3 was a control subject.

4.1.5 Data Analysis

A debriefing was realized with each pilot after the end of the twentieth and last landing to determine in particular whether:

- the audio alarm had been perceived or not;
- the cognitive countermeasures had helped the concerned pilots (*i.e.* Pilots 1 & 2) to take accurate decisions or not.

The performance of each pilot and the efficiency of the cognitive countermeasures were also assessed thanks to the objective analysis of the flight parameters recorded every 20 ms.

4.2 Results

4.2.1 Subjective Results for the Three Pilots

All the three pilots admitted they had not perceived the audio and visual alarms relative to the antiskid failure as shown in the second column of the next table. Pilots 1 and 2 also admitted that their decision to go around was influenced by the countermeasures (see third column of the next table). No cognitive countermeasure was sent to Pilot 3 (see the section Procedure): he performed a high-energy landing that led the aircraft to veer off the runway.

Pilot ID	Alarms perception	Go around
1	-	+
2	-	+
3	-	-

The next sections focus on the results of the first two pilots.

4.2.2 Objective Results: Pilot 1

Figures 3 and 4 show Pilot 1's behavior during the last landing when he faced the three events:

- the progressive windshear is triggered at time t = 102 and does not make Pilot 1 go around as he keeps on descending. On the contrary Pilot 1 first increases the thrust (from time t = 107 to t = 111) and then reduces it (from time t = 107 to t = 111) in order to deal with the windshear;

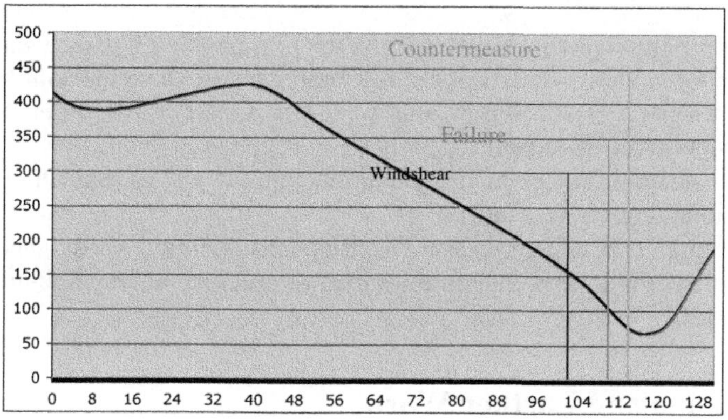

Fig. 3. this graph shows Pilot 1's altitude against time (in seconds) during his last landing. He faces the windshear (at t = 102, see the green vertical line), the failure (at t = 110, see the orange vertical line) and then the cognitive countermeasure (at t=114, see the blue vertical line).

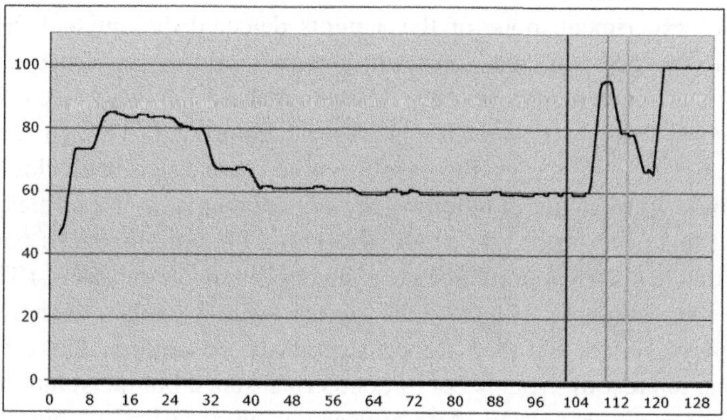

Fig. 4. This graph shows Pilot 1's throttle management against time (in seconds) during his last landing (0% corresponds to the "thrust idle" lever position, 100 % corresponds to the "go-around" lever position). He faces the windshear (at t = 102, see the green vertical line), the failure (at t = 110, see the orange vertical line) and then the cognitive countermeasure (at t=114, see the blue vertical line) that makes him go around.

- the antiskid failure is triggered at time t = 110 but Pilot1 maintains his descent;
- the cognitive countermeasure is triggered at time t = 114 and make the pilot abort his descent at time t = 115 and set the lever to maximum thrust in order to initiate a go around at time t = 119.

4.2.3 Objective Results: Pilot 2

Pilot 2's behavior during the last landing is close to Pilot 1's behavior when he faces the three events:

- the progressive windshear is triggered at time t = 83. This event does make him go around as he keeps on descending below the glide path from time t = 83 to time t = 106. From time t = 93 to time t = 108, Pilot 2 manages to recapture the glide path;
- the antiskid failure is triggered at time t = 110 but Pilot2 keeps on maintaining his descent;
- the cognitive countermeasure is triggered at time t = 123 and makes the pilot set the lever to maximum thrust in order to initiate a go-around at time t = 130.

5 Conclusion and Further Work

The aim of the exploratory experiments were respectively to define cognitive countermeasures and to initiate a consideration of the perseveration topic in the Airbus pilots community. The arguments are convincing: in the first experiment, the cognitive countermeasures were particularly efficient to cure pilots of perseveration; in the second experiment, none of the 3 pilots detected the antiskid failure and all persevered in a non-stabilized approach; the two pilots who faced the cognitive countermeasures reacted appropriately. Moreover the pilots who faced the cognitive countermeasures took 6 seconds to initiate the go-around. Though they were not prepared to such kind of alarms, this reaction time is consistent with classical reaction times of pilots facing an alarm requiring an immediate action. Nevertheless one must consider that this experiment was conducted with a limited number of participants on a simulator that has several limitations (i.e. no copilot, no autothrust…) Therefore, the results have to be analyzed with care: the limitations may explain the non detection of the failure by the pilots and their trend to persevere in landing. The experiment has confirmed the interest of the concept of countermeasures, but the solutions have to be further investigated: the tested countermeasures were efficient, but the proposed solution has quite surprised the pilots.

These results have led to conduct dozens of participative sessions with Airbus pilots to discuss about the necessity/importance of the concept of countermeasures, to identify typical "perseverogenic" situations and to refine the associated countermeasures in terms of design, triggering conditions, user's acceptability and training.

On account of the generic results, a series of experiments is planned with industry involvement to work on a specific case study so as to precisely:

- identify specific cognitive countermeasures for dedicated perseverogenic scenarios and assess their added-value in terms of safety and pilots' reactions;
- define the cognitive countermeasures with the users and the manufacturers in order to warrant the acceptability of the users and within the legal frame.

References

1. Dehais, F., Tessier, C., Chaudron, L.: GHOST: experimenting conflicts countermeasures in the pilot's activity. In: IJCAI 2003, Proceedings of the Eighteen International Joint Conference on Artificial Intelligence, Acapulco, Mexico (2003)
2. Evans, J.: Bias in human reasoning: causes and consequences. Lawrence Erlbaum Associates, Hove and London (1990)
3. Lesire, C., Tessier, C.: Particle Petri nets for aircraft procedure monitoring under uncertainty. In: Ciardo, G., Darondeau, P. (eds.) ICATPN 2005. LNCS, vol. 3536, pp. 329–348. Springer, Heidelberg (2005)
4. Sarter, N., Sarter, M.: Neuroergonomics: opportunities and challenges of merging cognitive neuroscience with cognitive ergonomics. Theoretical Issues in Ergonomics Science 4, 142–150 (2003)
5. Parasuraman, R., Rizzo, M.: Neuroergonomics: The Brain At Work. Oxford University Press, Oxford (2007)
6. Dehais, F., Goudou, A., Labedan, P., Duret, V., Colongo, C., Pastor, J.: Neuropsychological aspects of autopilot supervision. In: NPsyErg 2006, First European Workshop in Neuroergonomics, Toulouse, France (2006)
7. Festinger, L.: A theory of cognitive dissonance. Stanford University Press, CA (1957)
8. Milgram, S.: Obedience to authority: an experimental view. Harpercollins (1974)
9. Beauvois, J.L., Bungert, M., Mariette, P.: Forced compliance: commitment to compliance and commitment to activity. European Journal of Social Psychology 25, 17–26 (1995)
10. Rhoda, D., Pawlak, M.: An assessment of thunderstorm penetrations and deviations by commercial aircraft in the terminal area. Technical report, M.I.T. (1999)
11. BEA, The get-home-it is syndrome. Technical report (2002), `http://www.beafr.org/etudes/gethomeitis/gethomeitis.html`
12. BSV, Chapitre Ejection. Bulletin Sécurité des Vols. Technical Report (2007)
13. Goh, J., Wiegmann, D.A.: Human factors analysis of accidents involving visual flight rules flight into adverse weather. Aviation, Space, and Environmental Medicine 73, 817–822 (2002)
14. Orasanu, J., Martin, L., Davidson, J.: Cognitive and contextual factors in aviation accidents: decision errors. In: Salas, Klein, G.A. (eds.) Linking expertise and naturalistic decision making, pp. 209–225. Lawrence Erlbaum, Mahwah (2001)
15. Muthard, E.K., Wickens, C.D.: Change detection after preliminary flight decisions: Linking planning errors to biases in plan monitoring. In: Proceedings of the 46th Annual Meeting of the Human Factors and Ergonomics Society. Santa Monica (2002)
16. O'Hare, D., Smitheram, T.: « Pressing on » intro deteriorating conditions: an application of behavioral decision theory to pilot decision making. International Journal of Aviation Psychology 5(4), 351–370 (1995)
17. Causse, M., Dehais, F., Péran, P., Démonet, J.F., Sabatini, U., Pastor, J.: Monetary incentive provokes hazardous landing decision making by enhancing the activity of "emotional" neural pathways. In: 15th Annual Meeting of the Organization for Human Brain Mapping, San Francisco, Etats-Unis (2009)

18. Keyser, V.D., Woods, D.: Fixation errors in dynamics and complex systems. In: Systems Reliability Assessment, pp. 231–251. Kluwer Academic Press, Dordrecht (1990)
19. Pastor, J.: Cognitive performance modeling - Assessing reasoning with the EARTH methodology. In: COOP 2000, Sophia Antipolis (2000)
20. Miyake, A., Friedman, N., Emerson, M., Witzki, A., Howerter, A.: The unity and diversity of executive functions and their contribution to complex "frontal lobe" tasks: a latent variable analysis. Cognitive Psychology 41 (2000)
21. Halligan, P.W., Kischka, U., Marshall, J.C.: Handbook of Clinical Neuropsychology. Oxford University Press, Oxford (2004)
22. Simpson, J., Drevets, W., Snyder, Z., Gusnard, D., Raichle, M.: Emotion-induced changes in human medial prefrontal cortex: I. During cognitive task performance. PNAS 98(2) (2001a)
23. Simpson, J., Drevets, W., Snyder, Z., Gusnard, D., Raichle, M.: Emotion-induced changes in human medial prefrontal cortex: II. During anticipatory anxiety. PNAS 98(2) (2001b)
24. Posner, M.I., Cohen, Y.: Components of visual orienting. In: Bouma, H., Bouwhuis, D.G. (eds.) Attention and performance X, pp. 531–556. Erlbau, Hillsdale (1984)
25. Camus, J.F.: La psychologie cognitive de l'attention. Armand Colin (1996)

First Experimentation of the ErgoPNets Method Using Dynamic Modeling to Communicate Usability Evaluation Results

Stéphanie Bernonville[1,2,3,4], Christophe Kolski[1,2,3], Nicolas Leroy[4], and Marie-Catherine Beuscart-Zéphir[4]

[1] Univ Lille Nord de France, F-59000 Lille, France
[2] UVHC, LAMIH, F-59313 Valenciennes, France
[3] CNRS, UMR 8530, F-59313 Valenciennes, France
[4] EVALAB-EA 2694, Faculté de Médecine, 1 place de Verdun, F-59045 Lille, France
{sbernonville,nicolas.leroy-2,mcbeuscart}@univ-lille2.fr, christophe.kolski@univ-valenciennes.fr

Abstract. When a computer application is being designed or re-engineered, especially a user-centred application, communication between ergonomists and computer scientists is very important. However, the formalisms used to describe ergonomic problems and recommendations are often based on natural language. Consequently, the results of ergonomic evaluation can be poorly understood or interpreted by computer scientists. To remedy this problem, we propose a method, called ErgoPNets. The method creates common work support for both the ergonomists and the computer scientists working on the same project. Comprehensible for everyone, this support must provide an efficient tool that can be used by each person involved. ErgoPNets uses Petri nets to model Human-Computer Interaction (HCI) procedures and ergonomic criteria to model the ergonomic analysis. A first experimentation has been performed with designers/developers and academic researchers.

Keywords: Human-Computer Interaction (HCI), HCI modelling, usability problems, ergonomic criteria, petri nets, ErgoPNets Method, critical system.

1 Introduction

In a design or re-engineering project, the role of ergonomists is to evaluate with their own methods (e.g., ergonomic evaluations and/or user tests) mocks-up, prototypes or applications where some are more complex than others (i.e. healthcare domain). In this case, it is necessary to describe in a rigorous way the detected problems and recommendations to avoid computer system failure or malfunctions [1], [2], [3].

To proceed to the ergonomic evaluation, ergonomists can analyse user activity and/or base their evaluations, recommendations and justifications of their results on the experience gained during previous analyses. Obviously, the ergonomists and the computer scientists must work in close collaboration to successfully complete the project [4].

The original version of this chapter was revised: The copyright line was incorrect. This has been corrected. The Erratum to this chapter is available at DOI: 10.1007/978-3-642-11750-3_10

P. Palanque, J. Vanderdonckt, and M. Winckler (Eds.): HESSD 2009, LNCS 5962, pp. 81–95, 2010.

Following an activity analysis or an ergonomic evaluation, ergonomists often encounter problems in communicating their ideas, recommendations and results [5], especially when the information must be interpretated by the computer scientists. In this study, we concentrate on the communication of the results from an ergonomic evaluation of computer tools and present an example to illustrate this problem.

Ergonomists perform usability inspections—such as cognitive inspections (e.g., cognitive walkthrough) [6], evaluations of conformity to recommendations (e.g., guideline reviews) [7], evaluations of conformity to ergonomic dimensions (e.g., standards inspections) [8]—in order to detect ergonomic problems and then recommend actions to take to solve the problems. Figure 1 shows one type of form which gives problems and recommendations from a project on which several of the authors of the present article worked. This type of formalisation is practically always accompanied by a screenshot of the software display and, when necessary, by the ergonomists' model of the solution to the problem detected. The chart includes a "criteria" column showing the ergonomic criterion that corresponds to the problem detected, a "problem description" column containing a text explanation of the problem, a "consequences" column indicating the possible risks that the problem could cause, a "recommendations" column containing a text explanation of the ergonomists' suggestions and a "degree of gravity" column indicating the seriousness of the problem, which could range from one star (not serious) to four stars (extremely serious). This type of description is simple to read but can lead to comprehension and interpretation problems. The fact that the problem detected and the recommendation for solving it are both described in natural language can give rise to the following problems: the problem description may be ambiguous, forcing the computer scientist to read the text several times in order to understand the problem, and even then he/she might misinterpret the description; the recommendation may not provide a precise solution; or the problem and/or its solution may not be situated in terms of the overall system dynamics.

Criteria	**Problem Description**	**Consequences**	**Recommendations**	**Degree of gravity**
Consistency	Certain symbols do not mean the same thing on different screen pages; for instance: "A" for the stopping of an order and "A" for an absence of patient progress, "R" for reactivate and renew.	Risk of error/disturbance Increase in memory workload	Do not use the same symbol or word to mean different things; always use the same formulation for the same meaning.	**

Fig. 1. Example of a form used by ergonomists for reporting ergonomic problems and recommendations, (translation of a form given in [9])

This article proposes a possible solution for solving these problems of comprehension and interpretation by suggesting a precise location of the problems detected and recommendations made, directly onto procedures. Moreover, the ErgoPNets method, works to create a common work support [10], [11], [12] [13] [14]. This method makes it possible to model the ergonomic problems detected by the ergonomists, as well as their recommendations. ErgoPNets models incorporate the formal modelling method, *Petri nets* and the ergonomic criteria that are used in usability engineering [15], [16].

The first section presents the ErgoPNets method. The second section describes the experimentation of the method and provides the first results obtained. The last section offers our conclusion and our perspectives for research.

2 Presentation of the ErgoPNets Method

In this section, we present a detailed description of the ErgoPNets method, including both the Petri nets and the ergonomic criteria. We discuss the reasons which led us to choose the Petri nets formalism and the ergonomic criteria.

2.1 Petri Nets

Petri nets (PN) have been used in Human-Computer Interaction (HCI) for almost twenty years. They were first used to model human tasks [17] and [18], then they progressively came to be used in the specification and design of targeted interactive systems, particularly dynamic systems [19] [20] [21]; see for example articles about ICO (Interactive Cooperative Objects): [22], [18] and [23]. When linked to the object concept, Petri nets are used as a modelling tool in Task Object Oriented Design (TOOD) [24] [25], which aims to provide a method that covers the entire design process from task modelling to HCI parts generation. Palanque and his colleagues have proposed rule-based mechanisms for the automatic evaluation of PN-based models of interactive systems [26]. Ezzedine and Kolski [27] used Petri nets to study the functional tasks of a technical system in normal and abnormal situations, in order to facilitate the specification of interactive systems. Petri nets can also be used as a tool for formal comparisons of a set human task (theory) and the corresponding real task (practice) [28]. These diverse use possibilities make Petri nets a good choice for our application. We also chose to use Petri nets because, in addition to their varied uses, they make it possible to represent the task's dynamic dimension graphically. In our context, they also make it possible to model the procedures provided by the software and to associate these procedures to the ergonomists' recommendations.

2.2 The Ergonomic Criteria

Ergonomic inspection is commonly used to judge the conformity of computer interfaces to usability principles [29]. During such inspections, a small group of ergonomists examine the interfaces in detail in order to assess their conformity. Though some ergonomists base their judgements on experience and intuition alone, the application of certain basic rules, set out in the form of guidelines, is recommended [30]. Amongst the most commonly-used rule structures for ergonomic inspections, we chose one taken from the ergonomic criteria developed by Bastien and Scapin [31]. To develop these criteria, experimental results and recommendations were synthesized and translated into rules, which were then grouped together, creating 8 criteria and 13 sub-criteria [31] (Fig. 2). The results of an experimental study project by Bastien and Scapin [32] showed that these ergonomic criteria were more efficient than the dialogue standards ISO/DIS 9241-10 with respect to detecting ergonomic problems on user interfaces.

Main Criteria	Sub-criteria
1. Guidance	1.1 Prompting 1.2 Grouping/Distinction by location 1.3 Immediate feedback 1.4 Legibility
2. Workload	2.1 Brevity 2.2 Information density
3. Explicit control	3.1 Explicit user action 3.2 User control
4. Adaptability	4.1 Flexibility 4.2 User experience
5. Error Management	5.1 Error protection 5.2 Quality of error messages 5.3 Error correction
6. Consistency	
7. Significance of codes	
8. Compatibility	

Fig. 2. Classification of the ergonomic criteria and sub-criteria developed by Bastien and Scapin

The criteria described above were created to help HCI evaluators detect problems during ergonomic inspections; they also represent the main ergonomic dimensions according to which an interactive software programme may be specified or evaluated. We therefore use them to categorize problems detected using other methods, such as observation or user testing. Indeed, the criteria were specifically designed in order to be used by both human factors specialists and non-specialists [32].

2.3 Principles of the ErgoPNets Method

Basic stages of the ErgoPNets method. The ErgoPNets method has five stages which can be applicated after an ergonomic evaluation performed by ergonomists: context definition, current procedure description, problem identification and explanation, recommended procedure description, and recommendation identification and explanation.

1. *context definition:* Defining the context means characterizing the software analysed (which could mean providing a specific reference to a report mentioning the software's HCI specification) and identifying the user's objective. This helps to clarify the context.
2. *current procedure description:* In this stage, the procedure provided by the existing software, prototype or mock-up, and corresponding to the user objective identified in stage 1, is described with the help of Petri nets, indicating the user actions and the results of these user actions with the current software.
3. *problem identification and explanation:* In this stage, the ergonomic problem is identified and its specific place in the procedure described in stage 2 is situated. Textual explanations are provided to clarify the extent and possible consequences of the problem.

4. *recommended procedure description:* In this stage, Petri nets are also used to describe a procedure that integrates the ergonomists' recommendation, and perhaps to provide a mock-up of the recommended procedure. This new procedure may be completely different from the one described in stage 2 or may just be changed slightly.
5. *recommendation localization and explanation:* In this stage, the changes recommended by the ergonomist are situated in the procedure described in stage 4 and some textual explanations are provided to clarify the improvement brought about by the changes.

Stages two and four call for an adapted formalization using Petri Nets. System states or places (small circles) can be actions taken by the user or by the computer application, and transitions (small rectangles) are the events that allow movement from one state to another. The places and transitions are linked by arcs (arrows). Each place and each transition is described in words, using "and", "or" and "then", thus allowing several actions or several events to be represented. These words imply different things. For example, using the word "and" does not impose a specific order of actions/events; however, the word "then" does.

Stages three and five identify and locate the sets of places and transitions that correspond to the problem and the recommendation. The locations are represented with a dotted line framing the part of the procedure in which the problem/ recommendation is found. In stage three, the ergonomic criteria—represented by icons and text describing the problems—are applied. Each criterion (issued from the work of Bastien & Scapin) has its own icon (shown in Table 1).

Table 1. Icons that represent the ergonomic criteria

Ergonomic criteria	Icons	Icons explanation
Guidance		Image of a signpost indicating directions to show persons the way
Workload		Image of the brain allowing the human-being to work
Explicit control		Image of remote control buttons (reverse and forward) allowing an explicit control of a video recorder
Adaptability		Image of the belt adjustable to the waist of persons
Error managment		Image of a warning notice board to prevent danger
Consistency		Group of non consistency forms
Significance of codes		Question mark meaning incomprehension
compatibility		Image of pieces of puzzle can be assembled

In stage five, the sign “R” (for recommendation) is used to indicate the result of a recommendation. The text that describes the problems and the recommendations is located in a text zone between two procedures and includes the name of the criteria and, maybe, the sub-criteria. This text zone is linked to the icons by a black line. It is also possible to show the correspondence between the two procedures with dotted grey lines, which allows the differences between two procedures to be highlighted.

The amount of detail in the procedures has been adapted to each situation. For example, some events were simplified because they are not necessary to understand the ergonomic problems (asterisk: *). However, these simplifications are still shown in order to represent a complete and logical procedure that fulfils the initial objective. To represent the information that *must* be given in interactive systems, the word "obligatory" can be added between brackets. Finally, it is possible to add comments about any element of the model using a rectangle with a turned down corner.

The procedure models provided by the software and the procedures illustrated by the mock-ups provide a good support that could facilitate software design and re-engineering. Ergonomists can use them to represent their recommendations in a manner that can be more easily exploited by development teams. The graphic representations used in the ErgoPNets method are shown in Table 2.

Table 2. Different graphic representations used in ErgoPNets

Forms	Meaning
○	Place (system state)
▭	Transition (event)
———▶	Link between the state of the system and the event
	Dotted line used to frame the problems and recommendations on the Petri Net
- - - - - - - - - -	Correspondence between procedures
	Sign indicating a proposed recommendation (place on the criterion icon to identify recommendations on the Petri Net)
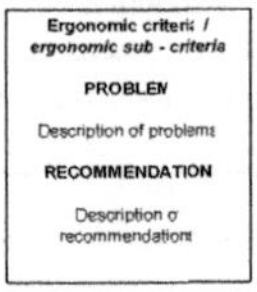	Text zone describing problems and recommendations
————	Link associating an icon with a text zone
*	Simplified procedure
AND OR THEN	The use of the words « and », « or » and « then » allows several actions or events to be represented. The word «and » does not impose an order, but the word « then » does.
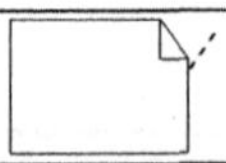	Note or comment
{obligatory}	Constraint indicating a compulsory event in the procedure

2.4 Tool Supporting the ErgoPNets Method

At the moment, the ErgoPNets models are built using the Visio© software. This software makes it possible to create plans and diagrams starting from predefined shapes. These shapes are organised in "template" files and are classified according to categories (e.g., software, flux diagram, network). To make a diagram, users simply need to choose the template they want and drag the shapes onto the drawing. Visio© makes it possible for users to create new templates and to define their own shapes. Using this software, we created an ErgoPNets template containing the various shapes needed to create a model. (All of these shapes have been shown above in Tables 1 and 2.) These are the graphic elements that allow Petri nets to be created and ergonomic problems and recommendations to be identified and described. Figure 3 shows an example of an ErgoPNets model created with the Visio© software. We are currently developing a software tool specifically for the ErgoPNets method. It is intended for use by both ergonomists and computer scientists. This software tool will allow the creation of common work supports to facilitate the modelling and analysis of the ergonomic problems detected, as well as the recommendations for solving these problems. In this version, this tool will not support the simulation of the models (see for instance PetShop supporting the simulation of ICOs [33]).

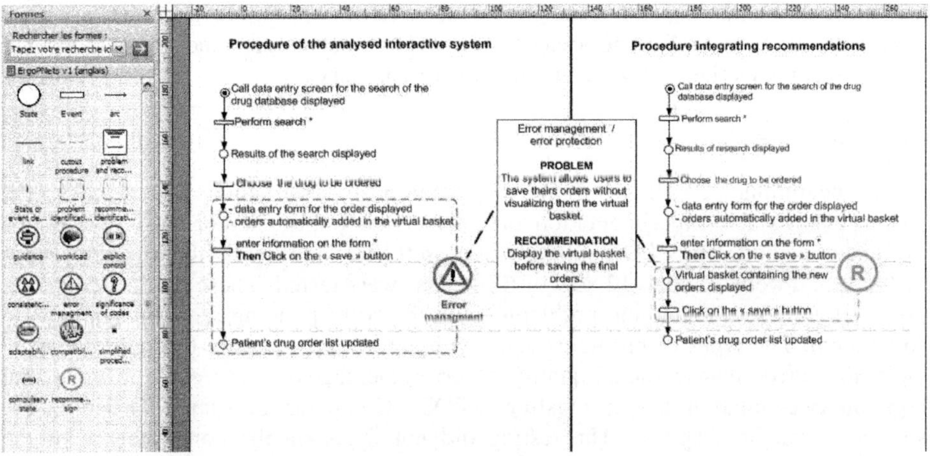

Fig. 3. Example of an ErgoPNets model created with the Visio© software

3 Experimentation of the ErgoPNets Method

The objective of the experimentation was to verify the following hypothesis: (1) a best comprehension level of problems and recommendations must be obtained with the ErgoPNets method compared with text based formalism, (2) the ErgoPNets is more relevant for problems linked to dynamic procedures.

For this, we asked a group of participants to estimate the comprehension of problems and recommendations described with the ErgoPNets method and with a based text formalism (an example of "table" formalism is described in figure 1) by comparing the final results of each method. The testing involved first reading a set of models described with the ErgoPNets method and then a second set of models described with the "table" formalism. This type of experimentation had the further advantage of allowing us to gather participant opinions about the ErgoPNets method.

3.1 Participants

Twelve participants were involved. Six designers/developers came from development companies who had worked on interactive software design projects. Their average age was 35 years, and they all had at least 5 years experience. This group was made up of people who used to work with ergonomists and therefore had analysed the descriptions of ergonomic problems and recommendations provided by the ergonomists. The second group included six academic researchers, specialists in software engineering and Human-Computer Interaction. Their average age was 30 years, and they all had at least 2 years experience. These people had a high level of knowledge about software design projects and about software engineering and HCI methods and models.

The objective was to test their understanding of the problems and recommendations described with the ErgoPNets method and the "table" formalism. The different nature of the two groups allowed us to collect information about ErgoPNets from two different points of view: (1) potential users and (2) methods and models specialists (i.e., software engineering researchers).

3.2 Procedure

At the beginning of each test some explanation and instruction was given to the participants as the experimentation objectives, the important concepts of the two evaluated methods, the CPOE software used in the test and finally the differents steps of the test. Twelve models (2 sets of 6 models) were tested. These models cover six potentially critical ergonomic problems, with the corresponding recommendations to solve them. The types of problems dealt with are shown in figure 6 (e.g., Guidance/ incitation, error management/ quality of error messages). They come from a real ergonomic evaluation of an existing CPOE (Computer Physician Order Entry) software used in hospitals. The testing did not focus on the completeness of the ergonomic criteria; indeed, the problems were chosen, based on typical problems associated to the procedures provided by the software. Examples of support used during the test are presented in fig. 4 and fig. 5.

Two groups have been planned for the evaluation. Each group was made up of three designers and three academic researchers. The first group tested problems numbered 1, 2 and 3 (see Fig. 6) described with the ErgoPNets method as well as problems numbered 4, 5 and 6 (see Fig. 6) described with "table" formalism. While the second group tested problems numbered 1, 2 and 3 (see Fig. 6) described with the "table" formalism as well as problems numbered 4, 5 and 6 (see Fig. 6) described with the ErgoPNets method. This organisation allowed facilitating the comparison of problems and their recommendations with the different participants.

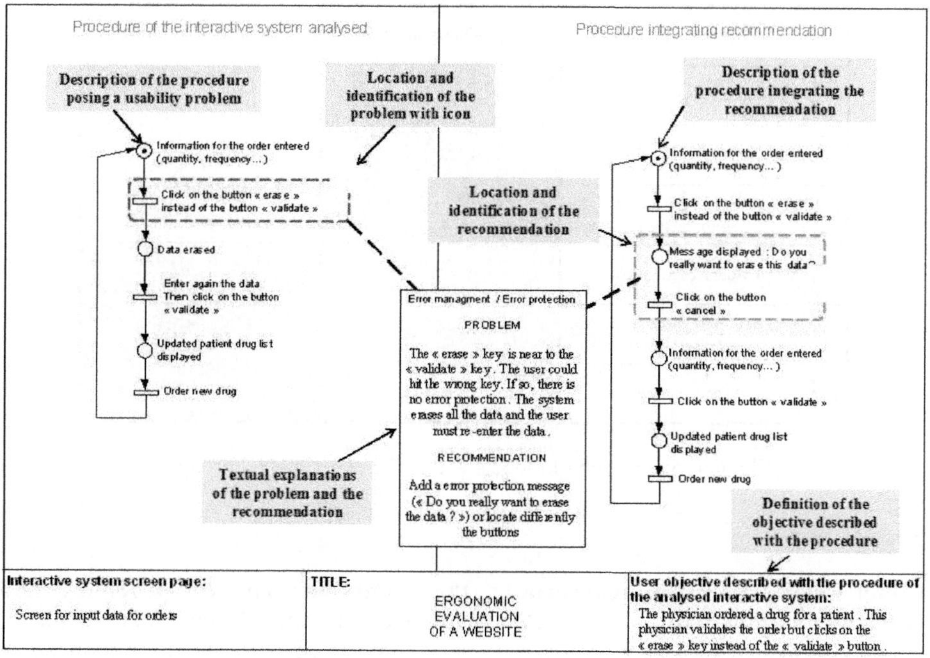

Fig. 4. Example of a model (n°5), described with the ErgoPNets method, involving the same problem and the same recommendation as the model in Figure 5 (please note that grey boxes are just comments for reader of this paper)

Criteria	Problem description	Consequences	Recommendations	Degree of gravity
Error managment/ error protection	The « erase » key is near to the « validate » key. The user could hit the wrong key. If so, there is no error protection. The system erases all the data and the user must re-enter the data.	Time loss	Add a error protection message (« Do you really want to erase the data? ») or locate the keys differently	✻✻

Fig. 5. Example of a model (n°5), described with the "table" formalism

Six questionnaires were given to the participants, who were asked to evaluate each model read. These questionnaires allowed us to collect data about the comprehension of the problems and recommendations described with each formalism: ErgoPNets or the "table" formalism. The participants must estimate their level of comprehension on a scale from 0 (complicated) to 10 (very clear). Finally, a global evaluation questionnaire was filled out, giving the opinion of the participants on the ErgoPNets method (difficulties encountered, deficiencies in the problem/recommendation descriptions, overall satisfaction/dissatisfaction).

Problems	ErgoPNets models	« Table » models
1. Guidance/incitation	GROUP 1	GROUP 2
2. Error managment/ quality of error messages		
3. Error managment/ error protection		
4. Adaptability/flexibility	GROUP 2	GROUP 1
5. Error management/ error protection		
6. Guidance/incitation		

Fig. 6. Organisation of models within the experimental protocol

3.3 Results

Comparison of results concerning the comprehension of problems and recommendations with the ErgoPNets method and the "table" formalism. To obtain the results presented below, we asked the participants their opinion about models of problem and models of recommendation. These indications were then compared to obtain information about the level of comprehension for 6 problems and their corresponding recommendations when using the ErgoPNets method and the "table" formalism. In general, the average scores obtained for problems modeled with the ErgoPNets method were superior to those obtained problems modeled with the "table" formalism, except for problems n°4 and n°6 (fig. 7, on the left).

For problem n°4, the software imposed a keyboarding order that was different from the usual physician ordering activity. The majority of the participants found that a textual explanation of the problem was sufficiently clear and that the ErgoPNets method was not necessary. They said they didn't need to understand the consequences of an action in a procedure.

For problem n°6, there was also a keyboarding problem. A textbox on a screen prompted the user to input data into this textbox which the software did not permit. Therefore, for the same reasons as in problem n°4, the majority of the participants found that a textual explanation was sufficient to describe the problem.

Our results also show that comprehending the problems and the recommendations was not easy for everyone, whatever the method used to describe the problem. Indeed, as the important differences on the graph in fig. 7 (on the left) show, the participants are divided. For example, for problem n°3 described with the ErgoPNets method, the scores were between 0 and 9.8. Two out of six participants gave a score between 0 and 5 (i.e., between *complicated* and *moderately clear*), while four out of six participants gave a score between 5 and 10 (i.e., between *moderately clear* and *very clear*). For problem n°3 described with the « table » formalism, the scores were positioned between 2 and 7.3. Four out of six participants gave a score between 0 and 5 (i.e., between *complicated* and *moderately clear*) and two out of six participants gave a score between 5 and 10 (between *moderately clear* and *very clear*). Thus, problem 3 was considered to be described more clearly with the ErgoPNets method, but the gap between the participant scores shows that these results are not unanimous.

The results for the recommendations were similar to those obtained for the problems. Globally, the recommendations were better understood if described with the ErgoPNets method, given the average scores obtained with the ErgoPNets method were superior to those obtained with the "table" formalism (fig. 7, on the right). The only exception was recommendation n°4, for which several participants felt that a simple textual description could be enough. Nonetheless, the differences between participants are important, underlining that the results do not show a clear advantage for ErgoPNets and those opinions are divided on the subject. In general, the participants liked the complete descriptions of the recommendations because, in their experience, the ergonomists' recommendations are often vague and open to interpretation.

In conclusion, these results are encouraging, but improvements still have to be made to obtain a clear mandate for ErgoPNets. The test has also helped to point out that the ErgoPNets method may be more effective for describing certain types of problems, especially for those associated with the procedure. The next section completes these results with a synthesis of all the information collected through the participant comments about the ErgoPNets method (e.g., participant satisfaction/dissatisfaction, lack of elements in the method proposed…).

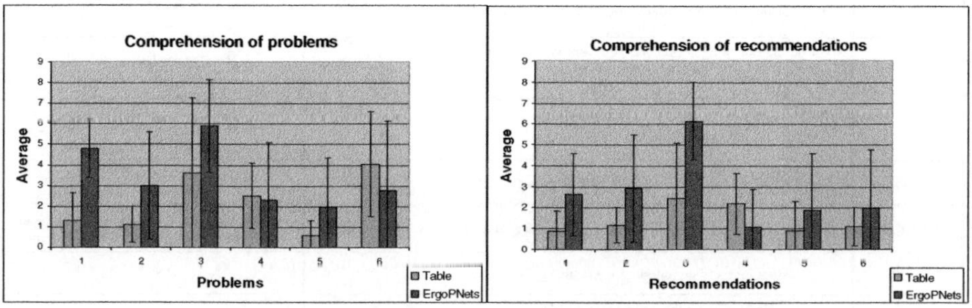

Fig. 7. Results obtained for the problems and the recommendations

Synthesis of the results based on the responses to global questionnaire and the comments recorded during the test. The global questionnaire and the comments recorded during the test allowed us to collect participant opinions about the ErgoPNets method. The questionnaire included questions about participant satisfaction, the difficulties encountered and the deficiencies of the ErgoPNets method. The comments of the designers and the developers were particularly useful, since they were able to assess the method's utility for their professional activity.

Table 3 reproduces some of comments recorded during the test. These comments were sorted into (1) positive points and the advantages of ErgoPNets and (2) negative points and disavantages of the method.

To measure participant satisfaction, we asked participants to indicate their level of satisfaction on a scale of 0 (not at all satisfied) to 10 (totally satisfied). The average score was 8.1 for the computer scientists, with a standard deviation equal to 1.60, and 7.8 for the researchers, with a standard deviation equal to 1.80. These results indicate a fairly positive view of the ErgoPNets method. Furthermore, five out of six computer-scientists answered *yes* to the question "Could the method ErgoPNets be helpful in your professional activity?".

In conclusion, all information collected allowed us, first, to assess the potential contribution of the ErgoPNets method for ergonomic evaluations of software within design or re-engineering projects. This method offers a promising way to describe the problems and recommendations encountered. In addition, the test also allowed us to identify the improvements that can be made (such as: addition of the degree of gravity). All of these remarks have been taken into account, and a new version of the ErgoPNets method is under development. When it is ready, a new test will be planned to evaluate the method's evolution.

Table 3. Synthesis of the comments recorded during the test (The same comment attributed to more than one subject means that their comments were very close.) (CS= Computer Scientist, R= Researcher).

Participants concerned	Positive points and advantages of the ErgoPNets method	Participants concerned	Negative points and disadvantages of the ErgoPnets method
CS2, CS6, CS7	Exhaustive and obvious representation of software procedures	CS5, CS6	It's still only « gymnastics for the mind »
CS2, CS7	The method allows a comparaison between the procedure integrating the problem and the procedure integrating the recommendation	R3, CS2	It could weigh down simple cases
R1	« The description of the procedures show one coherent result with all the project participants »	R1, CS7	Here the page is overloaded
CS2	«This type of method is less subject to interpretation»	CS6, CS7, R6	The model takes a long time to read
CS5, R1	This method helps to clarify things	CS7, R1	It can be complicated when there are several options for recommendations
CS6	«I find it easy because there is a logical sequence of actions»	R3	«However the method is not suitable for all that is only visual»
CS2, CS5	The model is enough to help developers working all alone	CS4	«Here I understood the ergonomic problem only after reading the recommendation»
R3	«ErgoPNets is appropriate for anything interactive»		
CS4	«Nothing better than diagrams to explain things»		
CS4	«We can see the different stages, It is better than a textual description where there is too much talk»		
CS5, CS7	The description allows the developer to see the result of a change in the procedure, it argues for the modification		

4 Conclusion and Perspectives

In this article, we presented a first experimentation concerning the ErgoPNets method helping to describe ergonomic problems and their recommendations. Its objective is to propose a formalism more rigorous than text-based description used currently in evaluation which can engender comprehension and interpretation problems. The ErgoPNets method combines Petri Nets and ergonomic criteria. This combination makes it possible to take two important aspects into account: (1) procedure descriptions and prescriptions and (2) HCI evaluation results, given the recommendations to be considered. The results of our tests of the ErgoPNets method show the method's potential for use in the evaluation phase of a design or re-engineering project. Even if they suggest further investigation, these results have allowed us to continue to improve certain aspects of the method. For example, we were able to improve the graphic representation in ErgoPNets by adding relevant elements and adapting the Petri nets to make the models easily understandable.

The ErgoPNets method is increasingly being used for real projects by ergonomists at the EVALAB Laboratory (see http://www3.univ-lille2.fr/evalab/).

For the moment, the tool supporting ErgoPNets is in the form of a Visio© template. We intend to develop this tool further, more specifically by allowing the verification of the Petri net properties. Such property verifications (e.g., network limits, network vivacity, network reversibility, or network blockage) will insure the formal character of the Petri nets created by ErgoPNets users and remove any potential incoherencies. At the moment, a first application has been developed. It takes into account the verification of rules linked to Petri net formalism. Our long term research perspective is to assemble, create and adapt software engineering methods and models in order to develop a multi-model approach that will facilitate the communication between the various partners in an information system design or re-engineering project.

Acknowledgements. The authors wish to thank the ergonomists and the computer scientists who participated in this study. They would also like to thank the Nord/Pas-de-Calais Regional Council, the CHRU of Lille, the FEDER, the RNTS network, and the French Ministry of Educ., Research and Technologies for supporting this research.

References

1. Horsky, J., Kuperman, G.J., Patel, V.L.: Comprehensive analysis of medication dosing error related to CPOE. JAMIA 12, 337–382 (2005)
2. Leveson, N.G., Turner, C.S.: An investigation of the Therac-25 accidents. Computer 26, 18–41 (1993)
3. Taylor, J.R.: The contribution of the design to accidents. Safety Sciences 45, 61–73 (2007)
4. Livari, N.: Representing the User' in software development—a cultural analysis of usability work in the product development context. Inter. With Computers 18, 635–664 (2006)
5. Gulliksen, J., Boivie, I., Göransson, B.: Usability professionals—current practices and future development. Interacting With Computers 18, 568–600 (2006)
6. Lewis, C., Wharton, C.: Cognitive Walkthroughs. In: Helander, M., Landauer, T.K. (eds.) Handbook of Human-Computer Interaction, 2nd edn., pp. 717–732. Elsevier Science, Amsterdam (1997)
7. Microsoft: The Windows Interface Guidelines for Software Design: An application Design Guide. Microsoft Press, Redmond (1995)
8. Shneiderman, B.: Designing the User Interface: Strategies for Effective Human-Computer Interaction, 2nd edn. Addison-Wesley Publishing Company, Massachusetts (1992)
9. Evalab, Ergonomic evaluation of the GENOIS software, Project Report, Evalab Laboratory Lille, France (2006)

10. Bernonville, S., Kolski, C., Beuscart-Zéphir, M.C.: Contribution and limits of UML models for task modelling in a complex organizational context: case studyin the healthcare domain. In: Soliman, K.S. (ed.) Internet and Information Technology in Modern Organizations: Challenges & Answers, Proc. 5th IBIMA Conf., Cairo, Egypt, pp. 119–127 (2005)
11. Bernonville, S., Kolski, C., Beuscart-Zéphir, M.C.: Towards an Assistance for Selecting Methods and Models for Interactive Software Design or Re-engineering within Complex Organisation: Application Case of a CPOE Software. In: Badr, Y., Chbeir, R., Pichappan, P. (eds.) Proc. of the second IEEE Int. Conference on Digital Information Management, Lyon, Workshop sessions, October 28-31, pp. 597–602. IEEE Press, Los Alamitos (2007)
12. Bernonville, S., Leroy, N., Kolski, C., Beuscart-Zéphir, M.C.: Explicit combination between Petri Nets and ergonomic criteria: basic principles of the ErgoPNets method. In: Proc. of the 25th Edition of EAM 2006, European Annual Conference on Human Decision-Making and Manual Control, Valenciennes, PUV, September 27-29 (2006) ISBN 2-905725-87-7
13. Beuscart-Zéphir, M.C., Pelayo, S., Guerlinger, S., Anceaux, F., Kulik, J.F., Meaux, J.J., Degoulet, P.: Computerized "Physician" Order Entry (CPOE): missing the "N", standing for Nurse. In: Physicians and Nurses activity analysis and comparison with Paper-based and Computerized Order Entry systems, IT in Health Care: Sociotechnical Approaches, Portland Oregon, USA (2004)
14. Bernonville, S., Kolski, C., Leroy, N., Beuscart-Zéphir, M.: Integrating the SE and HCI models in the human factors engineering cycle for re-engineering Computerized Physician Order Entry systems for medications: basic principles illustrated by a case study. International Journal of Medical Informatics, doi:10.1016/j.ijmedinf.2008.04.003 (in press)
15. Mayhew, D.J.: The usability engineering lifecycle. Morgan Kaufmann Publishers, San Francisco (1999)
16. Cockton, G., Lavery, D., Woolrych, A.: Inspection-based evaluations. In: Jacko, J.A., Sears, A. (eds.) Handbook of Task Analysis for Human-Computer Interaction, pp. 1118–1138. Lawrence Erlbaum Associates, London (2003)
17. Abed, M., Ezzedine, H.: Vers une démarche intégrée de conception-évaluation des systèmes Homme-Machine. Journal of Decision Systems 7, 147–175 (1998)
18. Palanque, P., Bastide, R.: Synergistic modelling of tasks, system and users using formal specification techniques. Interacting With Computers 9, 129–153 (1997)
19. Moussa, F., Riahi, M., Kolski, C., Moalla, M.: Interpreted Petri Nets used for human-machine dialogue specification. Integrated Computer-Aided Engineering 9, 87–98 (2002)
20. De Rosis, F., Pizzutilo, S., De Carolis, B.: Formal description an evaluation of user-adapted interfaces. International Journal of Human-Computer Studies 49, 95–120 (1998)
21. Ezzedine, H., Trabelsi, A., Kolski, C.: Modelling of an interactive system with an agent-based architecture using Petri nets, application of the method to the supervision of a transport system. Mathematics and Computers in Simulation 70, 358–376 (2006)
22. Palanque, P.: Modélisation par objets coopératifs interactifs d'interfaces homme-machines dirigées par l'utilisateur. Thèse de Doctorat, Université de Toulouse 1 (1992)
23. Bastide, R., Navarre, D., Palanque, P.: A tool-supported design framework for safety critical interactive systems. Interacting with Computers 15, 309–328 (2003)
24. Tabary, D., Abed, M.: A software environment task object oriented design (ETOOD). Journal of Systems and Software 60, 129–140 (2002)
25. Abed, M., Tabary, D., Kolski, C.: Using Formal Specification Techniques for the Modelling of Tasks and Generation of HCI Specifications. In: Diaper, D., Stanton, N. (eds.) The Handbook of Task Analysis for Human-Computer Interaction, vol. 5, pp. 503–529. Lawrence Erlbaum Associates, Mahwah (2003)

26. Palanque, P., Farenc, C., Bastide, R.: Embedding Ergonomic Rules As Generic Requirements in a Formal Development Process of Interactive Software. In: Proceedings of Interact 1999, pp. 408–416. IOS Press, Amsterdam (1999)
27. Ezzedine, H., Kolski, C.: Modelling of cognitive activity during normal and abnormal situations using Object Petri Nets, application to a supervision system. Cognitive, Technology and Work 7, 167–181 (2005)
28. Abed, M., Bernard, J.M., Angué, J.C.: Method for comparing task model and activity model. In: Proceedings 11th European annual conference Human Decision Making and Manual Control, Valenciennes, France (1992)
29. Kahn, J., Prail, A.: Formal Usability Inspection. In: Nielsen, J., Mack, R.L. (eds.) Usability inspection method, pp. 141–171. John Wiley & Son, New York (1993)
30. Vanderdonckt, J., Farenc, C. (eds.): Tools for working with guidelines. Springer, London (2000)
31. Bastien, J.M.C., Scapin, D.L.: Ergonomic Criteria for the Evaluation of Human-Computer Interfaces, Rapport technique INRIA, 156 (1993)
32. Bastien, J.M.C., Scapin, D.L., Leulier, C.: The ergonomic criteria and the ISO/DIS 9241-10 dialogue principles: a pilot comparison in an evaluation task. Interacting With Computers 11, 299–322 (1999)
33. Barboni, E., Navarre, D., Palanque, P., Bazalgette, D.: PetShop: A Model-Based Tool for the Formal Modelling and Simulation of Interactive Safety Critical Embedded Systems. In: Proceedings of HCI'Aero conference (Demonstration) (HCI'Aero 2006), Seattle, USA (September 2006)

Contextual Inquiry in Signal Boxes of a Railway Organization*

Joke Van Kerckhoven, Sabine Geldof, and Bart Vermeersch

Namahn, Grensstraat 21, 1210 Brussels, Belgium
{jvke,sg,bv}@namahn.com

Abstract. A number of selected field-study techniques have been validated in a case study in the domain of railway signal boxes. The context of this work is the endeavour of a human-centred design consultancy to acquire know-how on HCI methods for use in the design of safety-critical systems. The field studies were aimed at providing a clear overview of the work environment, tasks and cognitive load of the signallers and of possible bottlenecks in the current way of operating. Our approach for conducting field studies in supervisory control systems is based on ethnomethodology, situation awareness, and mental models represented as an abstraction hierarchy. For each of these methods, we discuss our approach, the result, and the applicability of the technique to future safety-critical system design projects.

Keywords: Safety-critical systems, ethnomethodology, patterns of cooperative interaction, situation awareness, goal-directed task analysis, mental models, abstraction hierarchy.

1 Introduction

As the practice of human-centred design gains more ground, also in the domain of safety-critical systems, it is important to know which methods for field studies can yield the most practically useful results in specific settings. Safety-critical systems have stringent requirements in terms of error avoidance, efficiency and risk management. Performance errors are serious because of their enormous impact on human health or the environment. Typical applications include traffic control, process supervision, critical care and emergency response.

As a human-centred design consultancy involved in designing for safety-critical systems, we want to increase our internal know-how on interaction design for these systems, i.e. specialist knowledge, techniques and expertise. We conducted a research project funded by the Institute for the Encouragement of Scientific Research and Innovation of Brussels (IRSIB)[1]. In the context of this project, we performed field studies in railway signal boxes [1]. This paper describes our experience in validating a

The original version of this chapter was revised: The copyright line was incorrect. This has been corrected. The Erratum to this chapter is available at DOI: 10.1007/978-3-642-11750-3_10

* The research described is funded by IWOIB/IRSIB, the Institute for the Encouragement of Scientific Research and Innovation of Brussels (project reference: RBC/06 R-144).

[1] More information about the activities of IWOIB/IRSIB can be found on http://www.iwoib.irisnet.be/index_en.htm

P. Palanque, J. Vanderdonckt, and M. Winckler (Eds.): HESSD 2009, LNCS 5962, pp. 96–106, 2010.

number of research techniques in a case study and the conclusions drawn from this activity. In section 2 we briefly describe the context of our work, i.e. the larger research project. Section 3 introduces the practical context: the case study in which we performed the field study. Sections 4, 5 and 6 describe each of the validated frameworks. Our conclusions are presented in section 7.

2 Research Project on Methodology for Safety-Critical Systems

The goal of the research project was to acquire know-how about selected Models, Theories and Frameworks [2] ('MTFs') that might be useful in interaction design for safety-critical systems and in translating that knowledge into practice-oriented and market-worthy methods. To this end, we adopted a three-step process and applied it to a number of MTFs.

2.1 Hypothesis

Candidate MTFs were assessed on their supposed relevance to the domain of safety-critical systems. Figure 1 positions the selected MTFs with regards to the social dimension (whether the individual user or the larger environment is concerned) and to the human-centred design process. The following MTFs were the object of our study: Ethnomethodology (ETHNO), Distributed Cognition (DC), Human Decision Making (HDM), Mental Modelling (MM), Cognitive Work Analysis and Ecological Interface Design (CWA/EID), Human Visual Perception (PERC), Motor Behaviour Models (MOTOR). For each MTF we explored existing literature and devised hypotheses about the applicability of the MTF. These hypotheses addressed the question "How could we use this theoretical knowledge when designing for safety-critical systems?"

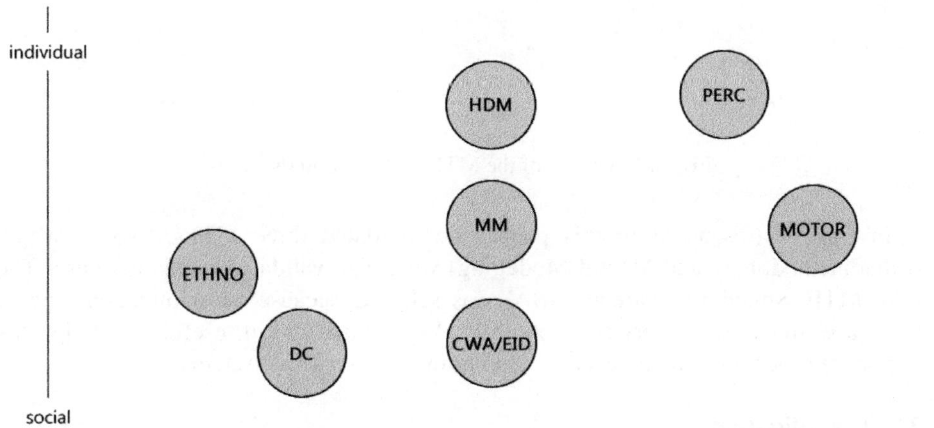

Fig. 1. Overview of the MTFs studied in the research project

2.2 Validation via Case Studies

Our understanding of these hypotheses and their value to the practice of safety-critical systems design were validated through realistic case studies in four different domains:

- Case 1: Diagnosis of liver cancer
- Case 2: Decision making for maxillofacial surgery
- Case 3: Configuration of ground-based satellite equipment
- Case 4: Supervisory control of railway signalling equipment (the subject of this paper)

Based on the outcome of the literature study and the type of case studies at hand, a number of hypotheses were assigned to each case, as shown in Figure 2.

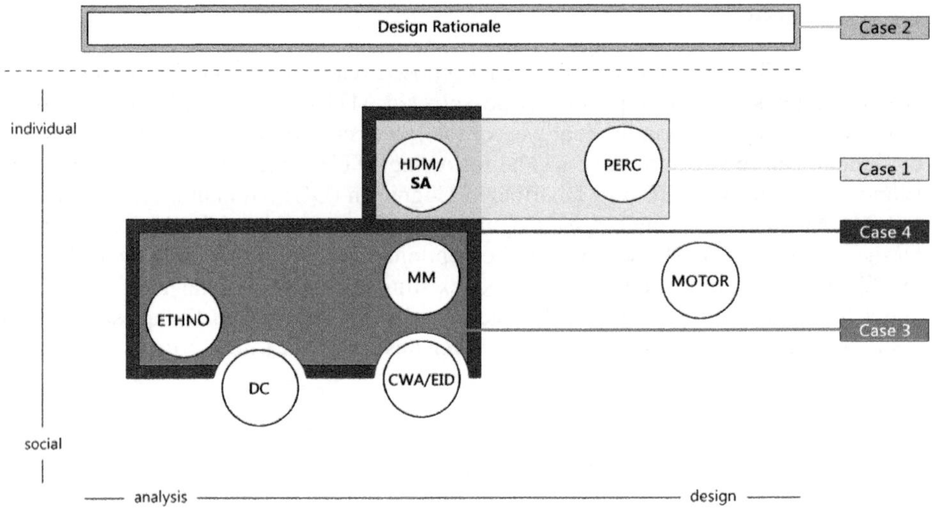

Fig. 2. Overview of the MTFs validated in the cases

In Case 4 (discussed in this paper) we validated three MTFs, two of which (Ethnomethodology and Mental Modelling) were also validated in another case. The third MTF, Situation Awareness (SA) was selected because the application type of this case study is 'supervisory control'. We considered it useful to study this important aspect of human decision making in safety-critical systems.

2.3 Consolidation

The lessons learned from the validation step were translated into methodology components. For each component, we provided a definition, a roadmap, resources and applicability criteria.

3 Case Study: Supervisory Control of Railway Signalling Equipment

This case study focused on the work of signallers in the signal boxes of a railway operator. From this control room, signallers manage the switches and signals of the railway infrastructure in a certain geographical area, comprising different action zones and sectors. The signal boxes we studied are equipped with a modern computer-based control system. Most routine jobs are automated, but interventions are required on a regular basis. The work is distributed as follows:

- An operator controls the signals and switches of a given action zone;
- A sector manager leads the operators working in his sector, which corresponds to a number of action zones;
- A zone regulator is in charge of the entire working zone of the signal box.

The workstations are grouped per sector and organized in two or three rows, as shown in Figure 3. In addition to displays on the desks, there is a large control display at the front of the room.

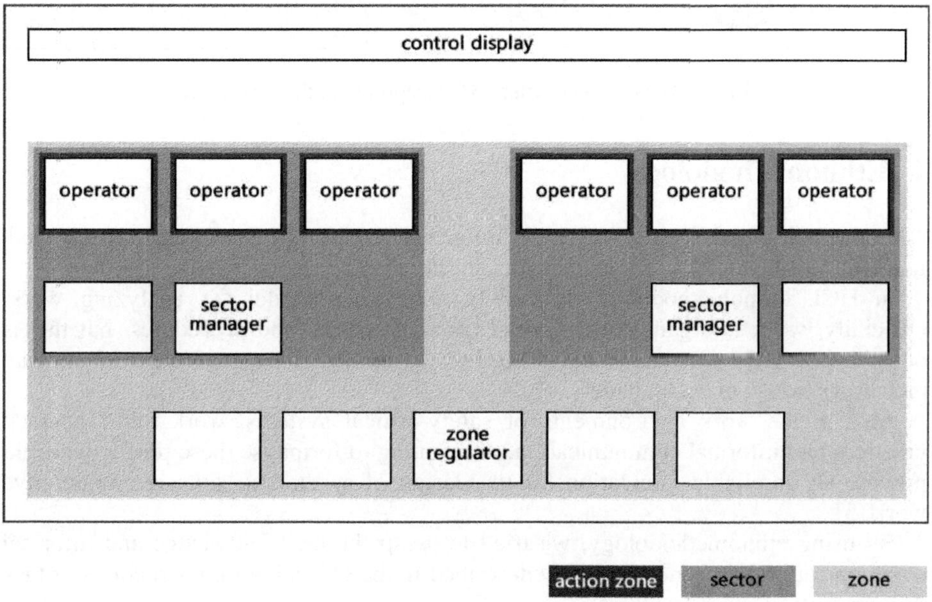

Fig. 3. Typical organization of a signal box

The aim of our research was to obtain a clear understanding of the cognitive and socio-technical dimensions of the work through field studies. So far, we have observed two operating signal boxes for five days in total. In order to structure and focus our field observations, we used the technique of ethnomethodology and insights from our previous study of situation awareness and mental modelling. As shown in

Figure 4, these MTFs complement each other in that they each deal with another aspect of the work:

- **Ethnomethodology** focuses on co-operation between the different actors in the control room.
- **Situation awareness** is the awareness a person needs to have about system events in order to make correct decisions.
- **Mental modelling** elicits the internalized image a user applies to mentally simulate and reason about the behaviour of a system.

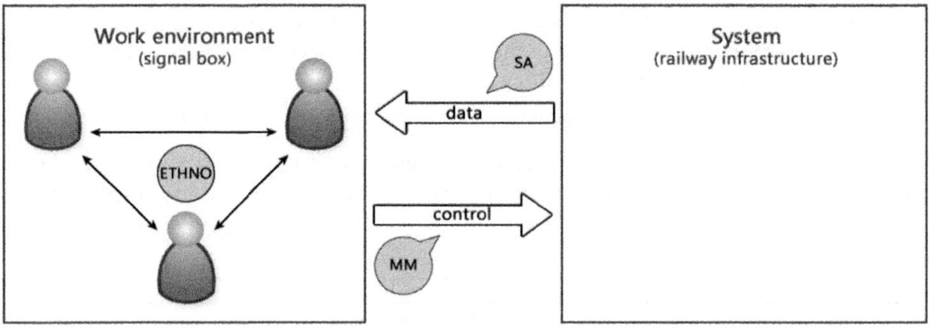

Fig. 4. The complementary MTFs applied in the case study

4 Ethnomethodology

Ethnomethodology is a sociological discipline that studies how social order is achieved. It assumes that people construct their social order in the process of acting together.

In HCI, ethnomethodology is mainly used as a model for analyzing work. Officially, work is organized by formal tasks, processes and procedures, but this is only in theory. In practice, workers need to get the job done, regardless of the way their work is organized on paper.

Also in the work environments of safety-critical systems, work relies on tacit practices and informal communication. Attempting to formalise these practices would provide an unreliable foundation for the design of systems to support co-operative work [3].

By using ethnomethodology, we tried to grasp the tacit knowledge and informal communication that is not explicitly described in the official work instructions or task descriptions.

4.1 Approach: Patterns of Cooperative Interaction

Traditional ethnomethodological analyses are time consuming. Several months are needed for observations, analysis, and interpretation. Furthermore, the results are specific to a certain working environment and certain situations, so they are not reusable.

However, Martin and Sommerville [4] identified and described 10 *patterns of cooperative interaction* from previous ethnomethodological studies that can be generalized and reused. The patterns are defined as 'regularities in the organisation of work, activity, and interaction amongst participants, and with, through and around artefacts'. The pattern descriptions are easily accessible, even for analysts or designers who lack the expertise of an ethnographer.

Most of these patterns are not specific to safety-critical systems, but they are recognized as important in field studies for systems design [4]. We used them as a focal point during our observations, and structured our findings according to them.

4.2 Result

Collaboration in small groups is one of the patterns we identified when observing the signallers' work. This pattern is concerned with the manner in which small, co-located groups collaborate to carry out tasks. The pattern draws attention to the way in which collaboration is facilitated by seating arrangements and various artefacts. [4]

In signal boxes, a sector manager and his operators cooperate intensively in order to be able to efficiently respond to the situation. They continuously shout to each other to exchange information quickly. The acoustics of the signal box is thus a crucial factor affecting the success of the collaboration.

We found that the usual seating arrangement in signal boxes is not ideal. There is no visual contact between a sector manager and his operators, which hinders communication. One is not able to visually verify whether the other party has successfully received a message. Furthermore, the sector manager is not able to see what the operators are doing.

The large control display at the front of the signal box is a typical example of the pattern *public artefact*. The display should facilitate communication, but unfortunately, it is hardly ever used. The information displayed shows insufficient detail, whereas a complete overview of the working zone of the signal box is necessary for decision making. In addition, the set-up of the signal boxes makes it impossible for signallers to see the full screen: the display doesn't fit into the field of vision from the first or second row of seats. Small font sizes make the information on the display illegible.

The patterns described above are only two examples of patterns we recognized. It turns out that all patterns listed by Martin and Sommerville were applicable to the signallers' work in some way.

4.3 Evaluation

The patterns of cooperative interaction provide a valuable framework for conducting field studies within a limited budget and timeframe. They offer a useful guide to structure observations themselves as well as the subsequent reporting.

We expected that using the patterns would later help us to optimally gear the design of an application to the context of use. However, making the link between analysis and design remains a creative step. The patterns don't include answers to the following questions:

- What are the design consequences if a particular pattern is observed?
- What is the relationship between the patterns, for example, can they compete?
- Are some patterns preferred over others?

- Do some patterns indicate a bad practice which should be eliminated?
- Are there strongly recommended patterns that should therefore be installed by design?

In addition, we discovered that it would be useful to enrich each pattern with focal points, pitfalls or guidelines that one should take into account during observation.

5 Situation Awareness

Endsley, Bolte and Jones [5] describe *Situation Awareness* (SA) as 'being aware of what is happening around you and understanding what that information means to you now and in the future'. In control rooms, situation awareness is a human skill that arises, amongst others, through the peripheral awareness of users interacting with a variety of information sources (e.g. monitoring screens, colleague interactions, interaction with other devices) [6]. In our study, we focused on eliciting the information that is relevant for performing a particular task.

In work situations, situation awareness is closely related to the goals and objectives associated with a specific job. It is especially important for effective decision making, and therefore a crucial factor in safety-critical systems.

We wanted to investigate how situation awareness can be promoted in the signallers' work situation.

5.1 Approach: Goal-Directed Task Analysis

We validated *goal-directed task analysis* (GDTA), as suggested by Endsley, Bolte and Jones [5], as a method for examining what information signallers need to perform their job. GDTA seeks to discover the user's information needs for optimally achieving his primary goals and making well-founded decisions.

Our research focused on the information needs of the sector manager:

- What are the goals a sector manager is trying to achieve?
- Which decisions need to be made for attaining these goals?
- What information would a sector manager *ideally* need to know for making these decisions?

These questions couldn't be answered directly by signallers, but had to be elicited via interviews and observations during real-time operations. The information gathered was organized into a hierarchy of goals, sub goals, decisions and situation awareness requirements.

5.2 Result

The overall goal of a sector manager was defined as follows:

- Guarantee the safety of railway traffic.
- Guarantee the timing of railway traffic.
- Guarantee the comfort of passengers.

One of the sub goals classified under this last goal is 'give priority to trains that are vital links for passenger traffic'. The related decision and situation awareness requirements are illustrated in Figure 5.

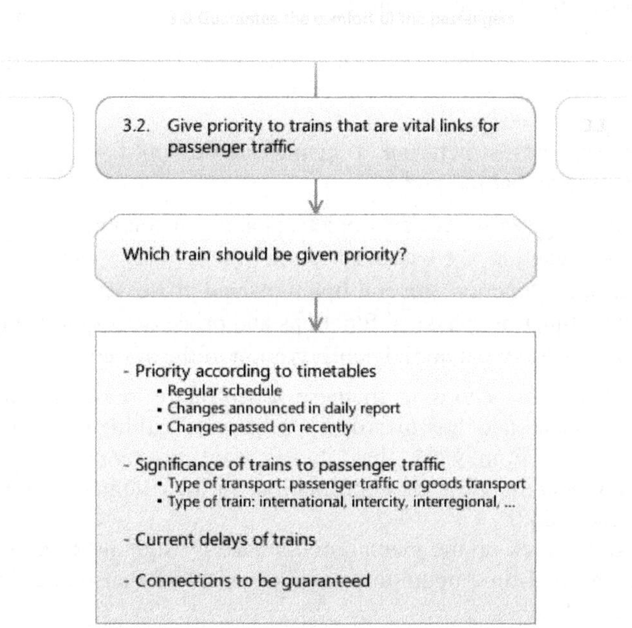

Fig. 5. Extract from the goal hierarchy of a sector manager

5.3 Evaluation

GDTA turned out to be a useful method to determine the information a user needs for making decisions. It is not easy to think in terms of goals instead of tasks, but in this case study, it proved to be a valuable method to uncover the information needs in a structured way, i.e. linked to the goal structure.

Furthermore, the focus on goals revealed an important difficulty which complicates the decision making process of a sector manager. It became clear that the three goals of the sector manager are contradictory. For example, raising the degree of safety has a negative effect on timing and vice versa. Therefore, a sector manager often struggles with making the right decisions: which goal should be given priority?

The given timeframe didn't allow us to create a complete goal hierarchy. In order to gain complete insight into the information needs, a more thorough analysis is required. This not only requires more observation time, but also evaluation of the analysis result (a GDTA structure) by users in several iterations.

6 Mental Models

To cope with problems arising in the control of complex systems, an operator has to address each problem at the appropriate level of cognitive resolution: some problems require a detailed view of the information, while for others, an overview is more appropriate so as to not make the problem unnecessarily complex.

We aimed to develop a mental model of the railway infrastructure that would help signallers structure the situation at several levels of complexity.

6.1 Approach: Abstraction Hierarchy

We wanted to verify whether the abstraction hierarchy, as proposed by Rasmussen [7], provides a good framework for representing a mental model of complex systems like a railway infrastructure.

Rasmussen suggests structuring a mental model on five levels of decreasing abstraction in a functional hierarchy:

- Functional purpose: system objectives and constraints
- Abstract function: the causal structure (physical laws, etc.)
- Generalized function: general functions and processes
- Physical function: physical functions and processes of the components
- Physical form: the material configuration of the system.

Structuring the system according to these levels helps reasoning about problems: upwards to a reason, and downwards to a cause. Within the domain of railway transportation, this framework has been used to create a macrocognitive representation model in view of coordinating various human factors projects and efforts within the field [8].

We attempted to pick up the mental model used by signallers via observations and interviews during real-time operations, and to visualize that model by means of an abstraction hierarchy.

6.2 Result

The result of our effort to create a layered mental model is shown in Figure 6. Note that this mental model is incomplete, because we encountered several difficulties as described below.

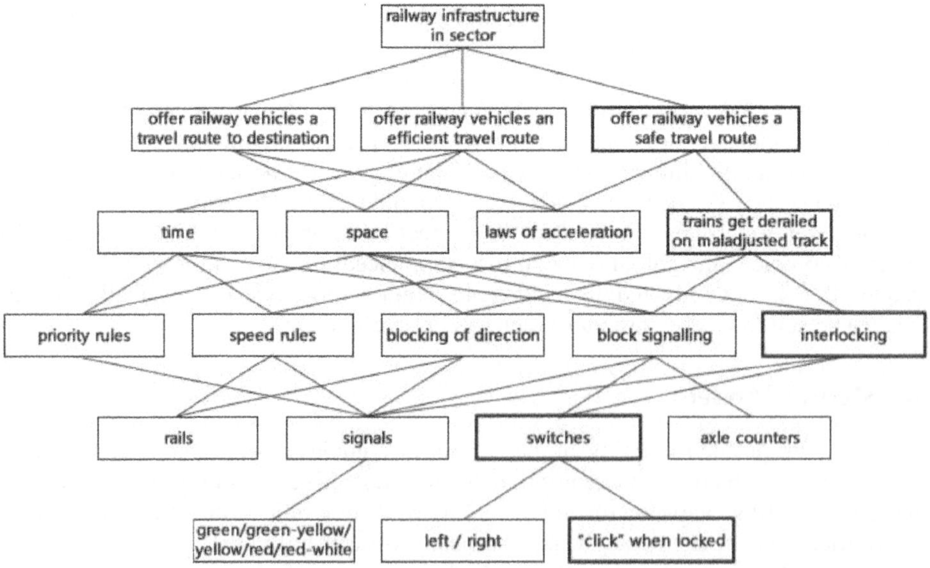

Fig. 6. (Incomplete) mental model of the railway infrastructure in five levels of abstraction

In the above diagram, several causal paths lead from the functional purpose of the system (top level) to a physical form (bottom level). Let's consider one of the paths as an example (the boxes with a darker outline).

'Offer railway vehicles a safe travel route' is one of the system objectives. To guarantee safety, one must take into account that trains can derail on a maladjusted track. 'Interlocking' is a typical process that ensures trains stay on the track: all switches on the desired route first have to be aligned and locked mechanically, before the train receives permission to depart. Switches are the physical components linked with interlocking. On a physical level, the locking of a switch manifests itself via an audible 'click'.

Following the path in the opposite direction, one can trace the link from a physical manifestation all the way up to a functional purpose of the system.

6.3 Evaluation

As human-centred designers, we feel the need to provide the user with an understandable system representation, which enables him to align his mental model with reality. However, we found that the mental model according to an abstraction hierarchy wasn't appropriate for this purpose.

Firstly, the mental model remains too abstract, because it lacks context and doesn't characterize important relationships, e.g. between the elements of one abstraction level. Also, in contrast with a traditional mental model, an abstraction hierarchy contains concepts about which a user is not or should not be aware. In this perspective, the abstraction hierarchy is more closely related to a business domain than a user's mental model. The result is that this mental model wouldn't succeed in supporting the user when interacting with the system.

Secondly, even though we have not yet completed the mental model, its complexity already has a degrading effect on the readability of the diagram. We suspect that completing the mental model would make this diagram unmanageably large and complex.

In addition, creating a good mental model representation according to an abstraction hierarchy is far from obvious. An in-depth understanding of the system is required. Since the mental model adds little value for communication with the users, the effort may not be justified.

For these reasons, we doubt the usefulness of the abstraction hierarchy for representing the mental model of signallers. However, others [8] found the framework useful in creating a high-level integrative model of railway operations to coordinate the efforts of a human factors team.

7 Conclusion

The aim of this project was to find out whether the selected MTFs are appropriate for field studies in safety-critical systems. Based on our field study in the signal boxes, we can draw the following conclusions:

- Patterns of cooperative interaction are useful for structuring observations. However, elaborating the specific characteristics for each pattern would significantly increase the added value.

- Goal-directed task analysis is a useful method to determine the correct information a user needs for making decisions, and to uncover conflicts between the goals one tries to achieve when making these decisions. However, we feel that user feedback on the resulting hierarchy is required.
- The added value of a mental model based on an abstraction hierarchy appeared small, certainly when compared to the effort it took to build it. Further investigation is needed to determine the circumstances in which an abstraction hierarchy would be appropriate.

We are already applying the know-how acquired through this research project to human-centred design projects for safety-critical systems.

With respect to our field study in the signal boxes, we hope to continue our analysis in order to elaborate our findings and offer useful advice for improving the working conditions of signallers.

References

1. Geldof, S., Van Kerckhoven, J.: Field study techniques for supervisory control systems. In: Drury, J. (ed.) Proceedings of the HCP-2008 Workshop on Supervisory Control in Critical Systems Management, Delft (2008)
2. Carroll, J.M. (ed.): HCI Models, Theories and Frameworks. Morgan Kaufmann, San Francisco (2003)
3. Heath, C., Luff, P.: Technology in action. Cambridge University Press, Cambridge (2000)
4. Martin, D., Sommerville, I.: Patterns of Cooperative Interaction: Linking Ethnomethodology and Design. ACM Transactions on Computer-Human Interaction 11(1), 59–89 (2003)
5. Endsley, M.R., Bolte, B., Jones, D.J.: Designing for situation awareness: An Approach to User-Centered Design. Lawrence Erlbaum Associates, NJ (2003)
6. Luff, P., Heath, C.: Naturalistic analysis of control room activities. In: Noyes, J., Bransby, M. (eds.) People in control. IEE control engineering series, pp. 151–167. The Institution of Electrical Engineers, London (2002)
7. Rasmussen, J.: The role of hierarchical knowledge representation in decision making and system management. IEEE Transactions on Systems, Man and Cybernetics 15, 234–243 (1985)
8. Bye, R., Farrington-Darby, T., Cox, G., Hockey, G.R.J., Wilson, J.R., Clarke, T.: Work Analysis and Distributed Cognition Representation of Integrated Rail Operations. In: Wilson, J., Norris, B., Clarke, T., Mills, A. (eds.) People and Rail Systems, pp. 275–283. Ashgate, Aldershot (2007)

Reducing Error in Safety Critical Health Care Delivery

Marilyn Sue Bogner

Institute for the Study of Human Error, LLC
msbogner@errorstudies.com

Abstract. A behavioral goal in the safety critical delivery of health care is minimizing the likelihood of error in the diagnosis and treatment of presenting symptoms. The prevailing belief in the United States is that information technology (IT) can effectively address such health care delivery problems. The veracity of this belief was not upheld by a study of the application of IT to health care (Stead & Lin, 2009). The study found that IT applications introduce problems such as difficult to detect forms of error and provide little support for clinicians' imposing cognitive tasks. This paper presents a model that addresses both of those issues. The model supports the clinicians' cognitive process of knowledge acquisition through differential diagnosis by creating an evidence based construct of the specific patient. The IT implementation of the model based on a systems engineering concept is described and implications considered for reducing the likelihood of error.

Keywords: Error, health care IT, cognitive support.

1 Introduction

More resources are spent on health care in the United States (U.S.) than in any other country yet the U.S. ranks last among the world's 30 industrialized countries. This is because a minority of the population receives excellent care and a somewhat equal minority receives adequate care (National Academies, 2006) a portion of the population that is approaching 50 million people essentially has no health care.

Pending the advent of re-designed health care to attempt to correct this situation, efforts are underway to reduce errors and enhance the quality of care by employing information technology (IT). Computer technology has great potential – it is easy to imagine how that technology can revolutionize the delivery of health care. The extent to which that currently is happening was addressed in a study by the Institute of Medicine (IOM) of the U.S. National Academy of Sciences involving eight medical centers that are exemplary in their application of IT in the delivery of health care (Stead & Lin, 2009).

1.1 The IOM Study of Reality

Based on observations and other data from site visits to the eight medical centers which included a Veterans Administration Center, the study committee reported that the applications of IT in health care did not differ appreciably from "bean counting"

The original version of this chapter was revised: The copyright line was incorrect. This has been corrected. The Erratum to this chapter is available at DOI: 10.1007/978-3-642-11750-3_10

P. Palanque, J. Vanderdonckt, and M. Winckler (Eds.): HESSD 2009, LNCS 5962, pp. 107–114, 2009.

which has been successful in administrative matters both in and out of health care. The committee also found that despite the potential of the technology, the health care IT in the subject medical centers rarely provided an integrative view of patient data important for avoiding errors in diagnosis and treatment (Stead & Lin, 2009).

Care providers were found to spend a great deal of time in electronically documenting what they did for patients. Clinicians often stated that such information was to comply with regulations or to defend against lawsuits, rather than information that someone could, indeed would use in improving patient care. The study committee found that IT exacerbated frustrations in performing some required tasks to the extent that work to accomplish those tasks increased rather than decreased. Another onerous finding was that often IT applications not only increased the possibility of error but also introduced new forms of error that are difficult to detect (Stead & Lin, 2009).

In summary, data gleaned from the visits to health care facilities considered exemplary in their implementation of IT suggest that if current applications health care IT continues unchanged, the nationwide implementation of IT will be insufficient to achieve the vision of 21st century health care – that health care should be safe, effective, patient-centered, timely, efficient, and equitable. More ominously, the study committee conjectured that the continuation of the current IT applications will set the goals of that vision of health care back to an earlier level (Stead & Lin, 2009).

The central conclusion of the IOM study provides a focus for change from the current unsatisfactory if not destructive course of IT health care applications. That focus is to provide IT support for the cognitive tasks inherent in the delivery of health care regardless of who provides it – family members, patients or professional clinicians. That support is envisioned as a cognitive model of the patient.

2 IT Cognitive Support in Health Care

Consistent with the "bean counting" approach of other aspects of health care IT applications, current IT based cognitive support typically focuses on health care transactions such as admitting a patient, interpreting a report, performing a procedure, (Stead & Lin, 2009). Such transactions involve activity to achieve a set of goals for the patient – goals that should be based on an understanding of the patient, yet raw data that are descriptive of the patient are illusive in current health care IT. Indeed, as previously noted clinicians spend considerable time searching for patient data that are vital to understanding the multiple interacting dimensions of the physiological, psychological and social aspects of the patient that is necessary to provide safe and effective care.

The emphasis on developing IT based cognitive support that is usable and useful for patient centered diagnosis and treatment of presents a formidable challenge to the computer science and biomedical informatics communities. That challenge can be met by gleaning insights from the study of human cognition by the discipline of psychology to enable an effective change from current IT applications.

An important insight for the IT support of cognitive tasks in health care is providing that support in a manner that is in harmony with the care providers health care related cognitive processes – processes that incorporate the specific provider's medical knowledge and experience. This is not as difficult as it may seem because clinicians are trained to acquire patient data by the process of differential diagnosis – a process that is amenable to IT implementation.

3 Differential Diagnosis

The process of differential diagnosis is the systematic acquisition of patient symptoms which often are characteristic of a variety of problems. This process of acquiring symptoms continues until one diagnosis can be differentiated from the various possibilities. Then treatment can be administered that is appropriate for the presenting problem. This differential diagnosis is guided by the knowledge and experience of the care provider. For example, the clinician acquires the datum that a patient is vomiting and also the datum that the patient has diarrhea. These data could lead to the diagnosis of flu or food poisoning. To clarify the issues for diagnosis, the clinician seeks additional data that might point to another source of the presenting symptoms.

Drawing on experience with patients having side effects from medication, the clinician finds that the patient is taking medication Z. That datum doesn't aid the diagnosis unless the care provider can access information concerning the side effects of Z from his or her experience or auxiliary information. Data are acquired that medication Z can have a side effect of vomiting and diarrhea. This additional datum changes the previous information/diagnosis so that rather than providing treatment for flu or food poisoning which would not have relieved the problem and possibly led to the supposition of error, the care provider discontinued medication Z, after which the vomiting and diarrhea stopped.

Although the above example is simple, the same cognitive processes occur in complex cases involving large numbers of variables – the more complex the case, the more important is the availability of support for the cognitive task of differential diagnosis. IT is ideal for this support – not because it is capable of complex, involved calculations which are not appropriate for differential diagnosis but because IT has the capability to support the simple yet challenging process of differentiating a myriad of data.

Thus, IT based support that simply presents data to the clinician in a manner amenable to his or her process of acquiring knowledge for differential diagnosis allows the clinician to form his or her conceptual construct of the specific patient. That construct can be elaborated by the acquisition of additional data to clarify the symptoms for effective diagnosis and subsequent treatment. The search for those additional data might be guided by the clinicians' previous experiences or a serendipitous datum which likely would not be present on a pre-determined patient model as in computer assisted diagnosis.

Because IT can present data so that they can be considered in different constellations the likelihood of an error occurring from not considering a datum relevant to the specific patient is reduced. It might be conjectured that using data to build a model for each patient is not an efficient use of time – an important consideration in this era of reimbursement dictated 12 minute patient encounters. This does not occur because the human mind and its cognitive processes are parsimonious. Representation of such parsimony that can guide the development of IT cognitive support is afforded by the Tien (2003) systems engineering framework of human cognitive structure. That framework supports the development of a patient centered cognitive construct for the integration of raw data into the context of the clinician's medical knowledge and experience that is described later in this paper.

4 A Framework of Human Cognitive Structure

The importance of a cognitive support model being compatible with the acquisition of data is underscored by the reference to cognition as the critical interface of humans interacting with data (Tien, 2003). A systems engineering elaboration of cognition (Tien, 2003; Tien & Goldschmidt, 2009) provides the basis for the ensuing discussion of the implementation of IT. That implementation which is related to the decision making framework in Figure 1 supports the process of differential diagnosis

The components of the framework vary in the dimensions of complexity with data being the least and wisdom the most complex, and time – data are the most transient and wisdom the least. Please note the tactical nature of the aspects of the framework indicated by the Decision Making Range area beneath of triangle in Figure 1 is beyond the scope of this discussion hence will not be addressed.

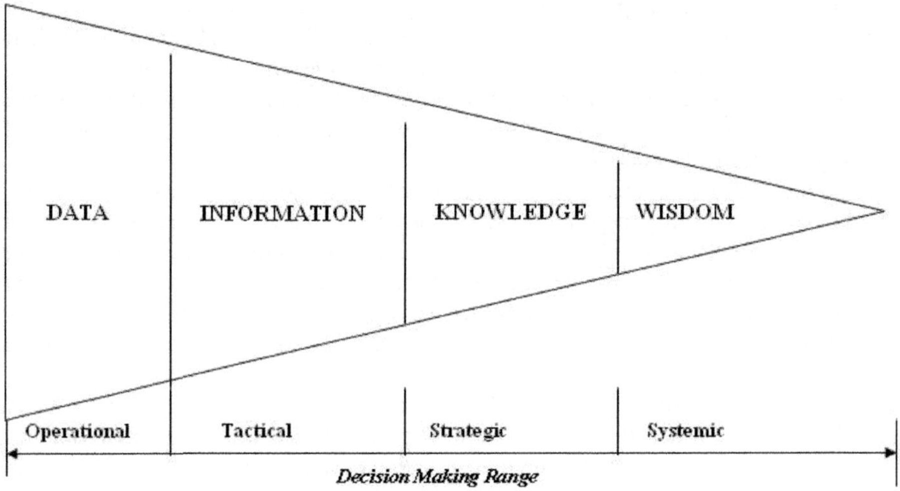

Fig. 1. Decision-Making Framework (Tien & Goldschmidt, 2008)

The components of the framework are not static or tightly compartmentalized. Rather their boundaries are permeable with elements moving within and across components. Each of the components of categories of elements in the framework is discussed in some detail to emphasize inter-relatedness among them. Such relatedness is relevant to the development of IT support for the process of differential diagnosis described later in the paper.

4.1 Data

Data, the basic elements of the decision framework, are obtained by observation, measurement, and verbal report such as a patient reporting a symptom. Data can be qualitative as well as quantitative. The patient's temperature and each aspect of the results from a laboratory test are examples of basic elements as are the symptoms of diarrhea and vomiting in the previous example.

Because humans are parsimonious, the individual datum is not absorbed by contributing to an information unit but continues as an element to be associated with other data to form other information units. It is emphasized that each datum is a single simple measure obtained at a given point in time – presenting compound material as data limits the clinician's ability to consider independently the elements that comprise the material for possible processing unless the clinician takes the time to decompose the material into its elements.

Each of multiple data elements of symptoms can be envisioned as cognitive elements that when acquired by the process of differential diagnosis and associated as in the previous example of diarrhea, vomiting, and use of medication Z are processed as a unit of information in the clinician's diagnostic cognitive structure.

4.2 Information

For each patient, data can be processed by linking each datum about that patient to an appropriate other – each elaborating the other. This continues until the data are processed by the clinician into information about the specific patient. Units of information can be characterized by patterns and groupings Although the data elements become information about patient A each datum is not absorbed by the processes but continues as an element in the care provider's cognitive structure being associated with other data elements to form information about a specific patient B or additional information about patient A.

Information includes groupings, and patterns, hence is more complex than data. Because the data comprising information were obtained at different times and the information is a part of the clinician's cognitive patient construct that might be used for future cases, it is considered analogous to short term memory in time (Wickens, et al, 2004)..

The process of data points becoming information continues in accordance with the education and experience of the clinician. The units of information may or may not be related in the clinician's cognitive structure. As more data points are obtained to elaborate the information or additional information is formed, appropriate and sufficient units of information coalesce to become knowledge.

4.3 Knowledge

The information units are not absorbed when they are inter-related and processed as knowledge – rather they are readily accessible so that other knowledge units might be formed by coalescing with different yet appropriate information units. Each data element that ties in to the information unit elaborates it and to the extent the information is an aspect of a knowledge, that knowledge is elaborated and can be accessible for the consideration of data points – the utility of data is multi-faceted.

The data points that lead to the decision then become information about the specific patient, which upon the existence of additional information about that patient can be processed when other conditions are met to become knowledge about the patient. The information components in a specific unit of knowledge are not absorbed in the knowledge but are available to consolidate as knowledge with other information units formed by the processing of when additional constellation of data points.

Because the data points for the specific patient have become information about that patient their usefulness as individual datum for that patient is diminished, hence their utility in the diagnostic process is transitory. This process occurs again for each specific patient often involving a number of considerations through differential diagnosis of data elements that provides more information for the clinician's cognitive construct of the specific patient.

Knowledge is more complex than the two previously noted aspects of the framework because it consists of synthesized information integrated with experiences, beliefs, values, attitudes, culture. Knowledge exists over time; it is analogous to long term memory (Wickens, et al, 2004).

4.4 Wisdom

Wisdom is synthesized knowledge plus insight often gleaned from multiple experiences and assessment of aspects of patient care over time. Units of knowledge regarding certain aspects of specific patients would be processed together with related experience based knowledge from a variety of sources. Wisdom is the most complex of the aspects of the cognitive framework, the most integral to the person, and not an aspect of every clinician's cognitive structure. Ideally a wise clinician would be encouraged to share to the extent possible the process by which the wisdom was acquired either verbally to the extent that one can articulate wisdom or by allowing unobtrusive observation of the application of that wisdom in patient care as in teaching.

The magnitude of each aspect of the framework is represented by the size of the triangle in Figure 1– the plethora of data being consolidated into a smaller quantity of information which in turn is synthesized to become a smaller quantity of knowledge, which is consolidated into wisdom.

5 IT Support for Cognitive Tasks

The previously discussed components of the human cognitive framework provide the operational concept for IT support for the cognitive tasks of health care diagnoses and treatment decisions. For ease of reference this IT cognitive support tool is referred to as Differential Diagnosis Cognitive Support (DDCS). It is to be noted that this is not an aid for diagnosing a specific health problem but support in diagnosing and treating the range of health problems. Neither is DDCS in any sense a computerized diagnosis. Rather DDCS serves as an extension of the care provider's patient specific cognitive processes – an extension that provides the raw material upon which the clinician can exercise his or her skills and knowledge to provide safe and effective patient centered health care.

The DDCS IT support is in the form of tree structure. The data and information the clinician enters is compared with data and information from the appropriate counterpart described in the discussion of the human cognitive structure. The clinician can select data or information at each level from one or more of three sources: that which the clinician previously entered that automatically entered the tree library, information from a data bank (from unidentified colleagues) or from the medical literature.

5.1 Data Entry

DDCS provides a diagnostic tree structure on which each datum of reported symptoms, observations and measurements for a specific patient would be entered individually. The clinician can prompt DDCS for an information tree with the same data elements together with unit(s) of information from those elements. That would appear on the screen. The clinician then can prompt for a tree with suggested additional categories of data elements that would aide in the process determining information to be used in considering the diagnosis for the specific patient.

If the clinician wishes guidance for gathering additional data for the specific patient, he or she can request that the DDCS data tree receive a thesaurus of DDCS data trees developed from the specific clinician's experience, or experiences of other unidentified clinicians for patients with demographic and health data comparable to those of the specific patient, or from the medical literature. Data for the specific patient would be entered which would trigger the presentation of several trees each with the entered data, suggestions for additional data to be acquired and the unit of information for each of those constellations of data. It may be necessary to enter more than one set of data trees depending on values entered for certain symptoms or results of tests.

When these data elements support information that the clinician considers sufficient for diagnosis, he or she might perform the differential diagnosis without IT assistance or request diagnosis trees tailored to the characteristics of the specific patient together with the medical information from the directory of diagnostic trees.

After forming a diagnosis appropriate for the specific patient, the clinician could request treatment trees from the directory of treatment trees. With the entry of appropriate patient characteristic data such trees could be tailored not only to the diagnosis but also to the extent possible to the personal characteristics of the specific patient to reduce the likelihood of error in treatment. Alternative trees can be requested

For those specific cases for which the treatment and its outcome potentially are problematic, the treatment trees for the specific patient can be submitted to the wisdom encyclopedia of outcomes.

6 Conclusion

The potency of this approach lies in data presentation that is driven not by what is technologically possible or by what researchers or theorists advocate rather by how the actual health care provider acquires and assimilates data. DDCS by incorporating information from actual clinical settings can provide insights gained from the real world experiences of clinicians. In addition the DDCS support of clinician's cognitive processes is dynamic. This is in contrast to the IOM study committee's vision of patient-centered cognitive support in which the clinician interacts with a static virtual patient comprised of raw data synthesized with medical knowledge in ways that make clinical sense for that patient.

The DDCS utilizes the capabilities of IT beyond "bean counting", but is DDCS feasible? That is an empirical question. Beyond a doubt there is a crying need for a technology such as DDCS that can reduce the likelihood of error by supporting the cognitive processing of the tsunamis of data for any given patient not only by clinicians but also by lay family care givers and patients involved in self care.

This need becomes critical when the stress of workload, production pressure, and fatigue compromise the cognitive functioning of the health care provider, hence increase the occurrence of errors. This need isn't unique to the U.S. – other countries have a similar issues perhaps to a lesser extent. Health is a precious commodity. It is time to realize the potential of IT to provide safe and effective patient centered care.

References

1. Stead, W., Lin, H. (eds.): Computational Technology for Effective Health Care: Immediate Steps and Strategic Directions. National Academies Press, Washington (2009)
2. Tien, J.: Towards a Decision Informatics Paradigm: A Real-Time, Information-Based Approach to Decision Making. IEEE Transactions on Systems, Man, and Cybernetics, Special Issue, Part C 33(1), 102–113 (2003)
3. Tien, J., Goldschmidt, P.: On Designing an Integrated and Adaptive Healthcare System. In: IOM Workshop: Engineering a Learning Healthcare System (2008)
4. Wickens, C., Lee, J., Liu, V., Gordon-Baker, S.: Introduction to Human Factors Engineering, 2nd edn. Prentice-Hall, Upper Saddle River (2004)

Erratum to: Human Error, Safety and Systems Development

Philippe Palanque[1], Jean Vanderdonckt[2], and Marco Winckler[1]

[1] University Paul Sabatier, Institute of Research in Informatics of Toulouse (IRIT), 118, Route de Narbonne, 31062 Toulouse Cedex 9, France
{palanque,winckler}@irit.fr

[2] Université Catholique de Louvain, Place des Doyens 1, 1348, Louvain-La-Neuve, Belgium
jean.vanderdonckt@uclouvain.be

Erratum to:
P. Palanque, J. Vanderdonckt, and M. Winckler (Eds.)
Human Error, Safety and Systems Development
DOI: 10.1007/978-3-642-11750-3

The book was inadvertently published with an incorrect name of the copyright holder. The name of the copyright holder for this book is: © Springer-Verlag Berlin Heidelberg. The book has been updated with the changes.

The updated original online version for this book can be found at DOI: 10.1007/978-3-642-11750-3

P. Palanque, J. Vanderdonckt, and M. Winckler (Eds.): HESSD 2009, LNCS 5962, p. E1, 2010.

Author Index

GPSR Compliance

The European Union's (EU) General Product Safety Regulation (GPSR) is a set of rules that requires consumer products to be safe and our obligations to ensure this.

If you have any concerns about our products, you can contact us on ProductSafety@springernature.com

In case Publisher is established outside the EU, the EU authorized representative is:

Springer Nature Customer Service Center GmbH
Europaplatz 3
69115 Heidelberg, Germany

Batch number: 10105422

Printed by Printforce, the Netherlands